ENGINEERING HYDROLOGY

Engineering Hydrology

E. M. WILSON, Ph.D., M.Sc., F.I.C.E., M.A.S.C.E.

Professor of Hydraulic Engineering
University of Salford

SECOND EDITION

First edition 1969
Second edition 1974, reprinted 1975, 1977, 1978, 1979, 1980

Published by
THE MACMILLAN PRESS LTD
London and Basingstoke
Associated companies in Delhi Dublin
Hong Kong Johannesburg Lagos Melbourne
New York Singapore and Tokyo

ISBN 0 333 17443 7

Printed in Hong Kong

70 12

P W
28/1/91

Contents

Preface to the Second Edition

This book has been written for engineering students and junior engineers who are learning about hydrology for the first time.

It is designed to introduce the reader to the elements of the subject and to the underlying theories, to show how these are modified in practice, and to describe the techniques which give answers in particular cases.

Most texts of this nature are American in origin and tend to be based exclusively on North American data and practice. This book still relies heavily on the American development of hydrology but also freely uses British and European data and references.

This edition has been revised and enlarged to provide more readily available information about rainfall in the British Isles, and in particular intensity–duration–frequency relationships. Recent work in the United Kingdom on consumptive use and infiltration has been incorporated and the section on stream gauging and flow measurement expanded. The opportunity has been taken to improve and update the references, to adopt SI units more uniformly throughout (though still not completely) and to correct many of the errors of the first edition. In this latter respect, I am particularly grateful to correspondents from many parts of the world.

To improve the book when used as an undergraduate text, a selection of problems has been given at the end of the main text, to allow readers to use the techniques described in the text.

I must record my appreciation of the advice and help received from many people during the book's preparation.

Manchester, 1974

E. M. WILSON

1 Introduction

The science of hydrology deals with the occurrence and movement of water on and over the surface of the earth. It deals with the various forms of moisture that occur, the transformation between the liquid, solid and gaseous states in the atmosphere and in the surface layers of land masses. It is concerned also with the sea: the source and store of all the water that activates life on this planet.

1.1 Allied Sciences

The engineer is normally concerned with the design and operation of engineering works to control the use of water and in particular the regulation of streams and rivers and the formation of storage reservoirs and irrigation canals. Nevertheless, he must be aware of this application of hydrology in the wider context of the subject, since much of the subject is derived from physics, meteorology, oceanography, geography, geology, hydraulics and kindred sciences. He must be aware of experience in forestry and agriculture and in botany and biology. He must know probability theory, some statistical methods and be able to use economic analysis.

Hydrology is basically an interpretive science. Experimental work is restricted, by the scale of natural events, to modest researches into particular effects. The fundamental requirement is observed and measured data on all aspects of precipitation, runoff, percolation, river flow, evaporation and so on. With this data and an insight into the many bordering fields of knowledge, the skilled hydrologist can provide the best solutions to many engineering problems that arise.

1

1.2 The hydrological cycle

The cyclic movement of water from the sea to the atmosphere and thence by precipitation to the earth, where it collects in streams and runs back to the sea, is referred to as the hydrological cycle. Such a cyclic order of events does occur but it is not so simple as that. First, the cycle may short-circuit at several stages, e.g. the precipitation may fall directly in the sea, in lakes or river courses. Secondly, there is no uniformity in the time a cycle takes. During droughts it may appear to have stopped altogether, during floods it

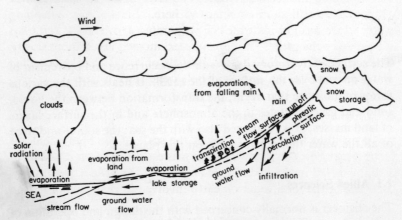

FIG. 1.1 *The hydrological cycle*

may seem to be continuous. Thirdly, the intensity and frequency of the cycle depends on geography and climate, since it operates as a result of solar radiation which varies according to latitude and season of the year. Finally, the various parts of the cycle can be quite complicated and man can exercise some control only on the last part, when the rain has fallen on the earth and is making its way back to the sea.

Although the concept of the hydrological cycle is oversimplified, it affords a means of illustrating the most important processes that the hydrologist must understand. The cycle is shown diagrammatically in Fig. 1.1.

Water in the sea evaporates under solar radiation and clouds of water vapour move over land areas. Precipitation occurs as snow, hail and rain over the land and water begins to flow back to the sea.

Some of it *infiltrates* into the soil and moves down or *percolates* into the saturated ground zone beneath the water-table, or phreatic surface. The water in this zone flows slowly through aquifers to river channels or sometimes directly to the sea. The water which infiltrates also feeds the surface plant life and some gets drawn up into this vegetation and *transpiration* takes place from leafy plant surfaces.

The water remaining on the surface partially evaporates back to vapour, but the bulk of it coalesces into streamlets and runs as surface runoff to the river channels. The river and lake surfaces also evaporate so still more is removed here. Finally, the remaining water which has not infiltrated or evaporated arrives back at the sea via the river channels. The *groundwater*, moving much more slowly, either emerges into the stream channels or arrives at the coastline and seeps into the sea, and the whole cycle starts again.

1.3 Inventory of earth's water

It is as well to have a clear idea of the scale of the events that are being discussed. The following table lists estimates of the amounts of water involved in the hydrological cycle and the proportion (in percentages) of the total water on earth involved in each part of it.

TABLE 1.1. *Estimated earth's water inventory*

Location	Volume 1000s of km^3	Percentage total water
Fresh-water lakes	125 ⎫	
Rivers	1·25 ⎪	0·62
Soil moisture	65 ⎬	
Groundwater	8,250 ⎭	
Saline lakes and inland seas	105	0·008
Atmosphere	13	0·001
Polar ice-caps, glaciers and snow	29,200	2·1
Seas and oceans	1,320,000	97·25
TOTAL	1,360,000 or $1·36 \times 10^{18}$ m^3	100·0

Of the 0·6% of total water which is available as fresh water about half is below a depth of 800 m and so is not practically available on the surface. This means that the stock of the earth's freshwater which is obtainable one way or another for man's use is about 4 million km^3 and is mainly in the ground. Spread over the earth's land surface it would be about 30 m deep.

The four processes with which the hydrologist is mainly concerned are precipitation, evaporation and transpiration, surface runoff or stream flow, and groundwater flow. He needs to be able to interpret data about these processes and to predict from his studies the most likely quantities involved in the extreme cases of flood and drought. He must also be able to express an opinion about the likely frequency with which such events will occur, since it is on the frequency of certain values of extreme events that much hydraulic engineering design is based.

1.4 Hydrology as applied in engineering

To the practising engineer concerned with the planning and building of hydraulic structures, hydrology is an indispensable tool. Suppose, for example, that a city wishes to increase or improve its water supply. The engineer first looks for sources of supply; having perhaps found a clear uninhabited mountain catchment area, he must make an estimate of its capability of supplying water. How much rain will fall on it? How long will dry periods be and what amount of storage will be necessary to even out the flow? How much of the runoff will be lost as evaporation and transpiration? Would a surface storage scheme be better than abstraction of the groundwater flow from wells nearer the city?

The questions do not stop there. If a dam is to be built, what capacity must the spillway have? What diameter should the supply pipelines be? Would afforestation of the catchment area be beneficial to the scheme or not?

To all these questions and many others which will arise the hydrologist can supply answers. Often they will be qualified answers and often also they will be given as probable values, with likely deviations in certain lengths of time. This is because hydrology is not an exact science. A contractor may be building a cofferdam in a river and his hydrologist may tell him that, built to a certain height, it will be overtopped only once, on average, in 100 years. If it is a temporary structure built for maybe 2 years' service only, the contractor may decide this is a fair risk. However, it *is* a risk. One of the 2 years may be that in which the once in 100 years flood arrives, and the science of hydrology cannot, as yet, predict this.

In a broader field of engineering, which is of great and increasing importance, the development of water resources over a whole river basin or geographical region may be under consideration. In these circumstances the role of the hydrologist is especially important. Now his views and experience are of critical weight not only in the engineering structures involved in water supply, but in the type and extent of the agriculture to be practised, in the siting of industry, in the size of population that may be supported, in the navigation of inland shipping, in port development and in the preservation of amenity.

Civilisation is primarily dependent on water supply. As the trend towards larger cities and increasing industrialisation continues, so will the role of the hydrologist increase in importance in meeting the demands of larger populations for water for drinking, irrigation, industry and power generation.

2 Meteorological Data

2.1 Weather and climate

The hydrology of a region depends primarily on its climate, secondly on its topography and its geology. Climate is largely dependent on the geographical position on the earth's surface. Climatic factors of importance are precipitation and its mode of occurrence, humidity, temperature and wind, all of which directly affect evaporation and transpiration.

Topography is important in its effect on precipitation and the occurrence of lakes, marshland and high and low rates of runoff. Geology is important because it influences topography and because the underlying rock of an area is the groundwater zone where the water which has infiltrated moves slowly through aquifers to the rivers and the sea.

The pattern of circulation in the atmosphere is complex. If the earth were a stationary uniform sphere, then there would be a simple circulation of atmosphere on that side of it nearest the sun. Warmed air would rise at the equator and move north and south at a high altitude, while cooler air moved in across the surface to replace it. The high warm air would cool and sink as it moved away from the equator until it returned to the surface layers when it would move back to the equator. The side of the earth remote from the sun would be uniformly dark and cold.

This simple pattern is upset by the earth's daily rotation, on its own axis, which gives alternate 12 hour heating and cooling and also produces the Coriolis force acting on airstreams moving towards or away from, the equator. It is further upset by the tilt of the earth's axis to the plane of its rotation around the sun, which gives rise to seasonal differences. Further effects are due to the different reflectivity and specific heats of land and water surfaces. The result of these circumstances on the weather is to make it generally complex and difficult to predict in the short term. By observations of data over a period of time, however, long term predictions may be made on a statistical basis.

The study of hydrology necessitates the collection of data, *inter alia*, on humidity, temperature, precipitation, radiation, and wind velocity. All of these are considered in this chapter.

2.2 Humidity

Air easily absorbs moisture in the form of water vapour. The amount absorbed depends on the temperature of the air and of the water. The greater the temperature the more water vapour the air can contain. The water vapour exerts a *partial pressure* usually measured in either *bars* (1 bar $= 10^5$ N/m^2; 1 millibar $= 10^2$ N/m^2) or mm height of a column of mercury (Hg). (1 mm Hg $= 1 \cdot 36$ mbar)

Suppose an evaporating surface of water is in a closed system and enveloped in air. If a source of heat energy is available to the system, evaporation of the water into the air will take place until a state of equilibrium is reached when the air is saturated with vapour and can absorb no more. The molecules of water vapour will then exert a pressure which is known as *saturation vapour pressure* or e_s, for the particular temperature of the system.

The value of e_s changes with temperature as indicated in Table 2.1. These values are also plotted as a curve connecting e_s and °C in Fig. 2.1. Referring to Fig. 2.1 consider what can happen to a mass of atmospheric air P, whose temperature is t and whose vapour pressure is e.

Since P lies below the saturation vapour pressure curve, it is clear that the air mass could absorb more water vapour and that if it did so while its temperature remained constant, then the position of P would move vertically up dashed line 1 until the air was saturated. The corresponding vapour pressure of P in this new position would be e_s. The increase $(e_s - e)$ is known as the *saturation deficit*.

Alternatively, if no change were to take place in the humidity of the air while it was cooled, then P would move horizontally to the

TABLE 2.1. *Saturation vapour pressure e_s in mm Hg (mercury) as a function of temperature t in °C (Negative values of t refer to conditions over ice)*

$$1 \text{ mm Hg} = 1\cdot36 \text{ mbar}$$

t	0·0	0·1	0·2	0·3	0·4	0·5	0·6	0·7	0·8	0·9	t
−10	2·15										−10
− 9	2·32	2·30	2·29	2·27	2·26	2·24	2·22	2·21	2·19	2·17	− 9
− 8	2·51	2·49	2·47	2·45	2·43	2·41	2·40	2·38	2·36	2·34	− 8
− 7	2·71	2·69	2·67	2·65	2·63	2·61	2·59	2·57	2·55	2·53	− 7
− 6	2·93	2·91	2·89	2·86	2·84	2·82	2·80	2·77	2·75	2·73	− 6
− 5	3·16	3·14	3·11	3·09	3·06	3·04	3·01	2·99	2·97	2·95	− 5
− 4	3·41	3·39	3·37	3·34	3·32	3·29	3·27	3·24	3·22	3·18	− 4
− 3	3·67	3·64	3·62	3·59	3·57	3·54	3·52	3·49	3·46	3·44	− 3
− 2	3·97	3·94	3·91	3·88	3·85	3·82	3·79	3·76	3·73	3·70	− 2
− 1	4·26	4·23	4·20	4·17	4·14	4·11	4·08	4·05	4·03	4·00	− 1
− 0	4·58	4·55	4·52	4·49	4·46	4·43	4·40	4·36	4·33	4·29	− 0
0	4·58	4·62	4·65	4·69	4·71	4·75	4·78	4·82	4·86	4·89	0
1	4·92	4·96	5·00	5·03	5·07	5·11	5·14	5·18	5·21	5·25	1
2	5·29	5·33	5·37	5·40	5·44	5·48	5·53	5·57	5·60	5·64	2
3	5·68	5·72	5·76	5·80	5·84	5·89	5·93	5·97	6·01	6·06	3
4	6·10	6·14	6·18	6·23	6·27	6·31	6·36	6·40	6·45	6·49	4
5	6·54	6·58	6·63	6·68	6·72	6·77	6·82	6·86	6·91	6·96	5
6	7·01	7·06	7·11	7·16	7·20	7·25	7·31	7·36	7·41	7·46	6
7	7·51	7·56	7·61	7·67	7·72	7·77	7·82	7·88	7·93	7·98	7
8	8·04	8·10	8·15	8·21	8·26	8·32	8·37	8·43	8·48	8·54	8
9	8·61	8·67	8·73	8·78	8·84	8·90	8·96	9·02	9·08	9·14	9
10	9·20	9·26	9·33	9·39	9·46	9·52	9·58	9·65	9·71	9·77	10
11	9·84	9·90	9·97	10·03	10·10	10·17	10·24	10·31	10·38	10·45	11
12	10·52	10·58	10·66	10·72	10·79	10·86	10·93	11·00	11·08	11·15	12
13	11·23	11·30	11·38	11·45	11·53	11·60	11·68	11·76	11·83	11·91	13
14	11·98	12·06	12·14	12·22	12·30	12·38	12·46	12·54	12·62	12·70	14
15	12·78	12·86	12·95	13·03	13·11	13·20	13·28	13·37	13·45	13·54	15
16	13·63	13·71	13·80	13·90	13·99	14·08	14·17	14·26	14·35	14·44	16
17	14·53	14·62	14·71	14·80	14·90	14·99	15·09	15·17	15·27	15·38	17
18	15·46	15·56	15·66	15·76	15·86	15·96	16·06	16·16	16·26	16·36	18
19	16·46	16·57	16·68	16·79	16·90	17·00	17·10	17·21	17·32	17·43	19
20	17·53	17·64	17·75	17·86	17·97	18·08	18·20	18·31	18·43	18·54	20
21	18·65	18·77	18·88	19·00	19·11	19·23	19·35	19·46	19·58	19·70	21
22	19·82	19·94	20·06	20·19	20·31	20·43	20·56	20·69	20·80	20·93	22
23	21·05	21·19	21·32	21·45	21·58	21·71	21·84	21·97	22·10	22·23	23
24	22·27	22·50	22·63	22·76	22·91	23·05	23·19	23·31	23·45	23·60	24
25	23·75	23·90	24·03	24·20	24·35	24·49	24·64	24·79	24·94	25·08	25
26	25·31	25·45	25·60	25·74	25·89	26·03	26·18	26·32	26·46	26·60	26
27	26·74	26·90	27·05	27·21	27·37	27·53	27·69	27·85	28·00	28·16	27
28	28·32	28·49	28·66	28·83	29·00	29·17	29·34	29·51	29·68	29·85	28
29	30·03	30·20	30·38	30·56	30·74	30·92	31·10	31·28	31·46	31·64	29
30	31·82	32·00	32·19	32·38	32·57	32·76	32·95	33·14	33·33	33·52	30
t	0·0	0·1	0·2	0·3	0·4	0·5	0·6	0·7	0·8	0·9	t

left along line 2 until the saturation line was intersected again. At this point P would be saturated, at a new temperature t_d, the *dewpoint*. Cooling of the air beyond this point would result in condensation or mist being formed.

If water is allowed to evaporate freely into the air mass, neither of the above two possibilities occurs. This is because the evaporation requires heat which is withdrawn from the air itself. This heat, called the *latent heat of evaporation*, h_r is given by the equation

$$h_r = 606\cdot5 - 0\cdot695t \text{ g cal/g} \tag{2.1}$$

So, as the humidity and vapour pressure rise, the temperature of the air falls and the point P moves diagonally along line 3 until saturation vapour pressure is reached at the point defined by e_w and t_w. This temperature t_w is called the *wet bulb temperature* and is the temperature to which the original air can be cooled by evaporating

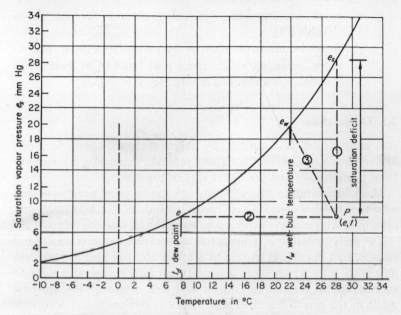

FIG. 2.1 *Saturation vapour pressure of water in air*

water into it. This is the temperature found by a wet bulb thermometer.

The *relative humidity* is now given as

$$h = e/e_s, \text{ or as a percentage, } h = 100\ e/e_s\% \qquad (2.2)$$

and is a measure of the air's capacity, at its existing temperature, to absorb further moisture. It is measured by blowing air over two thermometers, one with its bulb wrapped in wet muslin and one dry. The air flow past the bulb has an influence on the wet bulb reading and the two thermometers may either be whirled around on a string or more conveniently have the air current provided by a clockwork fan. In this latter case the instrument is called a *psychrometer*.

The value of e for air temperature t may be obtained from the equation,

$$(e_w - e) = \gamma(t - t_w) \tag{2.3}$$

where t_w = wet bulb temperature

t = dry bulb temperature

e_w = the corresponding partial pressures for t_w (from Table 2.1)

γ = psychrometer constant = 0·66 where e in mbar, t in °C and assuming an air speed past bulbs of at least 3 m/s (when e is measured in mm Hg, $\gamma = 0·485$)

2.3 Temperature

Air temperature is recorded by thermometers housed in open louvred boxes, about four feet above ground. Protection is necessary from precipitation and the direct rays of the sun.

Many temperature observations are made using *maximum and minimum thermometers*. These record by indices, the maximum and minimum temperatures experienced since the instrument was last set.

The daily variation in temperature varies from a minimum around sunrise, to a maximum from $\frac{1}{2}$ to 3 hours after the sun has reached its zenith, after which there is a continual fall through the night to sunrise again. Accordingly, max. and min. observations are best made in the period 8 a.m.–9 a.m. after the minimum has occurred.

The *mean daily temperature* is the average of the maximum and minimum and is normally within a degree of the true average as continuously recorded.

Temperature is measured in degrees Celsius, commonly, though erroneously, called centigrade. The Fahrenheit scale is also still used in Britain and almost exclusively in North America.

Vertical temperature gradient. The rate of change of temperature in the atmosphere with height is called the *lapse rate*. Its mean value is 6·5°C per 1000 m height increase. This rate is subject to variation, particularly near the surface which may become very warm by day, giving a higher lapse rate, and cooling by night giving a lower lapse rate. The cooling of the earth, by outward radiation, on clear nights may be such that a *temperature inversion* occurs with warmer air overlying the surface layer.

As altitude increases, barometric pressure decreases so that a unit mass of air occupies greater volume the higher it rises. The temperature change due to this decompression is about 10°C per 1000 m if the air is dry. This is the *dry-adiabatic* lapse rate. If the air is moist, then as it is lifted, expanding and cooling, its water vapour content condenses. This releases latent heat of condensation which prevents the air mass cooling as fast as dry air. The resulting *saturated-adiabatic* lapse rate is therefore lower, at about 5·6°C per 1000 m in the lower altitudes.

Distribution of temperature. Generally, the nearer the equator a place is the warmer it is. The effects of the different specific heats of earth and water, the patterns of oceanic and atmospheric currents, the seasons of the year, the topography, vegetation and altitude all tend to vary this general rule, and all need consideration.

2.4 Radiation

Most meteorological recording stations are equipped with *radiometers* to measure both incoming short-wave radiation from sun and sky and nett radiation which is the algebraic sum of all incoming radiation and the reflected short and long wave radiation from the earth's surface. The nett radiation is of great importance in evaporation studies as will be seen in Chapter 3.

2.5 Wind

Wind speed and direction are measured by *anemometer* and wind vane respectively. The conventional anemometer is the *cup anemometer* formed by a circlet of three (sometimes four) cups rotating around a vertical axis. The speed of rotation measures the wind speed and the total revolutions around the axis gives a measure of *wind run*, the distance a particular parcel of air is moving through in a specified time.

Because of the frictional effects of the ground or water surface over which the wind is blowing it is important to specify in any observation of wind, the height above ground at which it was taken. An empirical relationship between wind speed and height has been commonly used

$$u/u_0 = (z/z_0)^{0.15} \tag{2.4}$$

where u_0 = wind speed at anemometer at height z_0
$\qquad u$ = wind speed at some higher level z

In recent years there has been some effort to standardise observation heights and in Europe wind speed is usually observed 2 m above the surface.

Fig. 2.2 shows an instrument array for making meteorological observations at regular, short time-intervals. Instruments, which record automatically on magnetic tape, include nett-radiation radiometer, wet and dry bulb thermometers, wind vane, anemometer and incident solar radiometer at the mast-top.

2.6 Precipitation

The source of almost all our rainfall is the sea. Evaporation takes place from the oceans and water vapour is absorbed in the air streams moving across the sea's surface. The moisture-laden air keeps the water vapour absorbed until it cools to below dewpoint temperature when the vapour is precipitated as rain, or if the temperature is sufficiently low, as hail or snow.

The cause of the fall in temperature of an air mass may be due to convection, the warm moist air rising and cooling to form cloud and subsequently to precipitate rain. This is called *convective precipitation*. This is typified by the late afternoon thunderstorms which develop from day long heating of moist air, rising into towering anvil-shaped clouds. *Orographic precipitation* results from ocean air streams passing over land and being deflected upward by coastal mountains, thus cooling below saturation temperature and spilling moisture. Most orographic rain is deposited on the windward slopes. The third general classification of rainfall is *cyclonic and frontal precipitation*. When low pressure areas exist air tends to move into them from surrounding areas and in so doing displaces low pressure air upward, to cool and precipitate rain. Frontal rain is associated with the boundaries of air masses where one mass is colder than the other and so intrudes a cool wedge under it, raising the warm air to form clouds and rain. The slope of these frontal wedges can be quite flat and so rain areas associated with fronts may be very large.

2.6.1 Recording precipitation. Precipitation occurs mainly as rain, but may occur as hail, sleet, snow, fog or dew. Britain has a humid

Fig. 2.2 *Meteorological observation array. On lower arm l. nett radiation; r. wet and dry bulb thermometers; on upper arm l. wind direction; r. wind run: at top, solar and sky radiation. There is a rain gauge with anti-splash screen in the middle distance*

climate and rain provides the great bulk of its moisture, but in other parts of the world, precipitation may be almost entirely snow, or, in arid zones, dew.

In the United Kingdom rainfall records are received and recorded by the Meteorological Office from some 6500 rain gauges scattered over Gt. Britain and N. Ireland, the majority giving daily values of rainfall. In addition there are a further 260 stations also equipped with recording rain gauges which record continuously.

Standard rain gauges in Britain are made from copper and consist of a 5-inch diameter copper cylinder, with a chamfered upper edge, which collects the rain and allows it to drain through a funnel into a removable container of metal or glass from which the rain may be poured into a graduated glass measuring-cylinder each day. There are prescribed patterns for the standard gauge and for its installation and operation.

Recording gauges (or autographic rain recorders) usually work by having a clockwork-driven drum carrying a graph on which a pen records either the total weight of container plus water collected, or a series of blips made each time a small container of known capacity spills its contents. Such gauges are more expensive and more liable to error but may be the only kind possible for remote, rarely visited sites. They have the great advantage that they indicate *intensity* of rainfall which is a factor of importance in many problems. For this reason some stations are equipped with both standard and recording gauges.

The Meteorological Office has recently designed a new range of rain gauges.[1] The new standard gauge for daily rainfall measure is a circular catchment of 15×10^3 mm^2 (5·5 in. diameter) installed at a rim height of 300 mm above ground level. The larger and more accurate new gauge has an area of 75×10^3 mm^2 (12·2 in. diameter) also set with a rim 300 mm above ground. The material used in manufacture is fibreglass. New tipping-bucket mechanism has been designed and is available with a telemetry system to provide for distant reading by telephone interrogation. The gauge is provided with a telephone connection and number, which can be dialled in the ordinary way. The quantity of rain collected since the last setting of the gauge to zero is transmitted in increments of 1 mm by three groups of audible tones representing hundreds, tens and units. Interrogation can be made as frequently as desired and hence intensities can be obtained, by simple subtraction, with minimum delay.[2]

In recent years there has been much research into the effects of exposure on rain gauges and it is now generally accepted that more accurate results will be obtained from a rain gauge set with its rim at ground level, than one with its rim some height above ground [3]. It is necessary in a ground-level installation to make a pit to house the gauge and cover it with an anti-splash grid. Accordingly the ground-level gauge is more expensive to install and maintain.

A typical installation is illustrated in Fig. 2.3.

Yearly records for the whole country, statistically analysed and presented graphically, are published annually by the Meteorological Office in a booklet entitled *British Rainfall* followed by the particular year concerned. The use of rainfall data is discussed in Section 2.8.

2.6.2 Rain-gauge networks. A question which frequently arises is that concerning the number and type of rain gauges which are necessary to ensure that an accurate assessment of a catchment's rainfall is obtained. Bleasdale [4] quotes Tables 2.2 and 2.3 as general guides and comments as follows

The disparity between the two tables is not so great as might appear at first sight. The first indicates station densities which are reached in important reservoired areas and which may well be exceeded in small experimental areas. The second indicates densities which are more appropriate for country-wide networks. In the application of the general

TABLE 2.2. *Minimum numbers of rain gauges required in reservoired moorland areas*

mile²	km²	Daily	Monthly	Total
0·8	2	1	2	3
1·6	4	2	4	6
7·8	20	3	7	10
15·6	41	4	11	15
31·3	81	5	15	20
46·9	122	6	19	25
62·5	162	8	22	30

Source: Jnt Comm. Met. Off., R. Met. Soc. and Instn Wat. Engrs Rpt on "The determination of the general rainfall over any area". Trans. Instn Wat. Engrs, **42** (1937) p. 231.

FIG. 2.3 *Automatic recording rain gauge set with rim at ground level and with anti-splash screen*

guidance embodied in this Table [2.3] it must be appreciated that any large river basin will almost invariably have within it a number of sub-basins for which the relatively dense networks would be recommended. Moreover, the minimum densities suggested would often be substantially increased in mountainous areas, and would be closely followed only in areas of low or moderate elevation without complex topography.

TABLE 2.3. *Minimum numbers of rain gauges for monthly percentage-of-average rainfall estimates*

Area		Number of rain gauges
mile^2	km^2	
10	26	2
100	260	6
500	1300	12
1000	2600	15
2000	5200	20
3000	7800	24

Source: Bleasdale [4]

There is considerable material devoted to this question of hydro-logical network design and the reader is referred to the further references in the General Bibliography (p. 218).

2.7 Forms of precipitation other than rain

Snow and ice. Snow has the capacity to retain water and so acts as a form of storage. Its density and, therefore, the quantity of water contained, varies from as little as 0·005 for newly-fallen snow to as much as 0·6 in old, highly compressed snow. Its density varies with depth and so samples must be taken at various horizons in a snow pack before the water equivalent can be computed. This is usually done by sampling tube.

Snowfall may be measured directly by an ordinary rain gauge fitted with a heating system, or by a simple snow stake, if there is no drifting and density is determined simultaneously.

Snow *traverses* are made as field surveys along lines across catchments, to determine snow thicknesses and densities at depth so that water equivalents may be calculated for flood forecasts.

Fog. Estimates of amounts of moisture reaching the ground from fog formation have been made by installing fog collectors over standard rain gauges. Collectors consist of wire gauge cylinders on which moisture droplets form and run down into the rain gauge. Comparison with standard rain gauge records at the same locality shows differences which are a measure of fog precipitation. The interpretation of such data requires experience and the use of conversion factors, but can make substantial differences (of the order of 50–100%) to precipitation in forest areas.

Dew. Dew collectors have been used in Sweden and Israel to measure dew fall. They have been made as conical steel funnels, plastic coated and with a projected plan area of about 1 square metre. Dew ponds are used as a source of water in some countries. They are simply shallow depressions in the earth lined with ceramic tiles.

Condensation. Although fog and dew are condensation effects, condensation also produces precipitation from humid air flows over ice sheets and in temperate climates by condensation in the upper layers of soil. Such precipitation does not occur in large amounts but may be sufficient to sustain plant life.

2.8 The extension and interpretation of data

2.8.1 Definitions. The total annual amount of rain falling at a point is the usual basic precipitation figure available. For many purposes, however, this is not adequate and information may be required under any or all of the following headings:

(i) *Intensity.* This is a measure of the quantity of rain falling in a given time, e.g. mm per hour.
(ii) *Duration.* This is the period of time during which rain falls.
(iii) *Frequency.* This refers to the expectation that a given depth of rainfall will fall in a given time. Such an amount may be equalled or exceeded in a given number of days or years.
(iv) *Areal extent.* This concerns the area over which a point's rainfall can be held to apply.

2.8.2 Intensity–duration relationship. The greater the intensity of rainfall, in general, the shorter length of time it continues. A formula expressing the connection would be of the type

$$i = \frac{a}{t + b} \qquad (2.5)$$

where i = intensity: mm/h, t = time:h, a and b are locality constants, and for durations greater than two hours

$$i = \frac{c}{t^n} \qquad (2.6)$$

where c and n are locality constants.

The world's highest recorded intensities are of the order of 30 mm (or $1\frac{1}{4}$ in.) in a minute, 200 mm (or 8 in.) in 20 min and 26,000 mm (or 1000 in.) in a year. Further information is given in Section 9.6.

2.8.3 Intensity–duration–frequency relationships. In 1935 E. G. Bilham published his well-known article on these relationships in the United Kingdom [5] containing a graph reproduced here as Fig. 2.4. This graph used the subjective phraseology of "very rare", "remarkable" and "noteworthy" rather than frequency of occurrence. However, the frequencies were calculable from the formula

$$n = 1\cdot25t(r + 0\cdot1)^{-3\cdot55} \qquad (2.7)$$

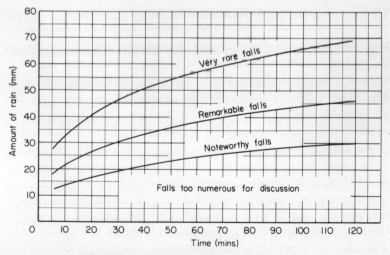

FIG. 2.4 *Bilham's rainfall classification*

where n = number of occurrences in 10 years
      r = depth of rain in inches
      t = duration of rain in hours
The SI form of the formula is

$$n = 1\cdot214 \times 10^5 t(P + 2\cdot54)^{-3\cdot55} \qquad (2.8)$$

where P = depth of rain in mm, and n and t have unchanged units.

In hydraulic engineering, however, it is usually the recurrence interval for a given depth of rain in a given time which is wanted. This is obtained by rearranging Eq. (2.8), so that

$$P = \left(\frac{(1\cdot214 \times 10^5)NT}{600}\right)^{1/3\cdot55} - 2\cdot54 \qquad (2.9)$$

where N = occurrence frequency, now expressed as once in N
      years = $10/n$
      T = duration in minutes

If intensities are required, the depth P is divided by the time in hours, i.e.

$$i = \frac{60P}{T} \text{ mm/h} \qquad (2.10)$$

and substituting the value of P from (2.9)

$$i = \frac{60}{T}\left((202\cdot3NT)^{-3\cdot55} - 2\cdot54\right) \text{ mm/h} \qquad (2.11)$$

Equation (2.11) is the SI form of the Bilham equation used in reference [6].

Bilham's work was revised and extended by Holland [7] who showed that Bilham's equations over-estimate the probabilities of high-intensity rainfall, i.e. above about 35 mm/h. This later work is best illustrated graphically and Fig. 2.5 shows both Bilham's formulae (chain-dotted lines) and Holland's revisions (full lines). The figure gives a return period for specific depths of rain occurring in specific periods of time, as averaged over 14 station-decades in England.

Another way of presenting such data, this time based on the correlation between annual average rainfall in Britain and one-day maximum rain-depth for various return periods is shown in Fig. 2.6.

For a specific locality it is often possible to produce curves such as those shown in Fig. 2.7 for Oxford, England, and Kumasi, Ghana,

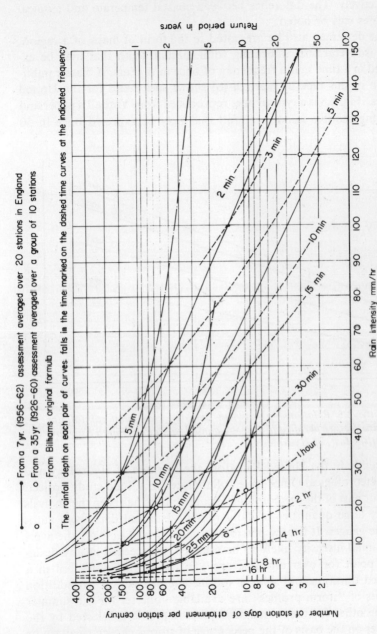

FIG. 2.5 *Rain-intensity–frequency graph (after D. J. Holland: "British Rainfall 1961", Crown Copyright). Example: A rainfall of 20 mm falling in a 30 min period may be expected, anywhere in Britain, on average once in 11.1 yr or 9 times a century.*

respectively. The difference between coastal temperate and tropical climates may be noted.

The data may also be presented in the form of maps of a region, with isohyetal lines indicating total rainfall depth that may be expected in a time t, at a frequency of once in N years. A classic publication of this type by Yarnall [8] shows such maps for the United States. Fig. 2.8 is a typical one, reproduced from Yarnall's paper and showing the five minute rainfall that may be expected once in 50 years.

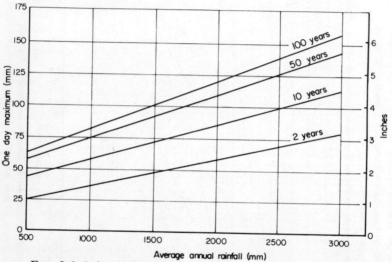

FIG. 2.6 *Relationship between one day maximum rainfall for given return period and average annual rainfall in United Kingdom (after Inst. of Hydrology)*

Recently in the United Kingdom the Institute of Hydrology and the Meteorological Office have collaborated in a flood study project. A major part of the meteorological contribution has been the analysis of very large quantities of daily rainfall data and data from autographic gauges. It is now possible to determine objectively the average rainfall intensity likely to be attained or exceeded at any designated point (or over any size area) in the United Kingdom in a specified duration and with a specified return period. In addition two sets of "storm profiles" are available—one set for the summer and the other for the winter half-year. A profile is selected by the engineer on the basis of the percentage of storms chosen to attain or exceed a given degree of profile "peakedness". A designer may apply a chosen rate of rainfall and a duration to a selected profile with a knowledge of the return period of the composite phenomenon. The latter information is of particular relevance to storm sewer design.

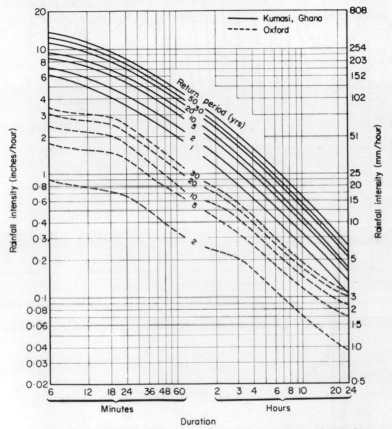

FIG. 2.7 *Rainfall frequency–intensity–duration relationships taken at Kumasi, Ghana (courtesy of Ghanaian Meteorological Service) and Oxford (courtesy of Inst. of Hydrology)*

2.8.4 Depth–area–time relationships. Precipitation rarely occurs uniformly over an area. Variations in intensity and total depth of fall occur from the centres to the peripheries of storms [9]. The form of variation is illustrated in Fig. 2.9, which shows for a particular

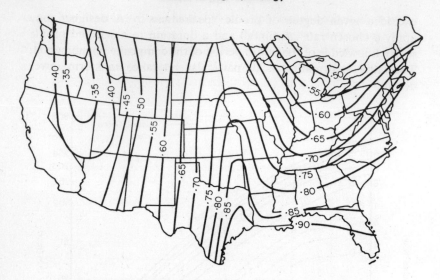

FIG. 2.8 *Five minute rainfall, in inches, to be expected once in 50 years in continental U.S.A. (after Yarnall)*

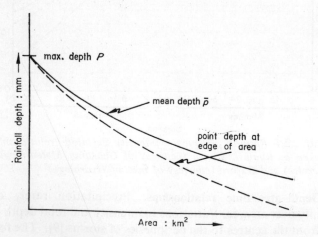

FIG. 2.9 *Depth–area curves of rainfall*

storm, how the average depth of fall decreases from the maximum as the considered area increases.

It is useful, however, to quantify this and Holland has shown [10] that the ratio between point and areal rainfall over areas up to 10 km² and for storms lasting from 2 to 120 minutes is given by

$$\frac{\bar{P}}{P} = 1 - \frac{0 \cdot 3 \sqrt{A}}{t^*} \qquad (2.12)$$

where $\bar{P}$ = average rain depth over the area

P = point rain depth measured at the centre of the area

A = the area in km²

t^* = an "inverse gamma" function of storm time obtained from the correlation in Fig. 2.10.

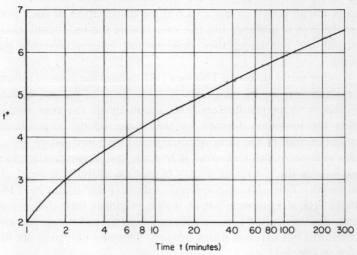

Time t (minutes)

FIG. 2.10 *Correlation between storm t and t^* (see ref. [11])*

Example 2.1. *What is the average rainfall intensity over an area of 5 km² during a 60-minute storm with a frequency of once in 10 years, in Britain?*

From Fig. 2.5, the frequency line of once in 10 years just cuts the 25 mm depth at about 1 hour ∴ $P = 25$ mm. Given $t = 60$ min then from Fig. 2.10, $t^* = 5 \cdot 6$ then

$$\frac{\bar{P}}{P} = 1 - \frac{0 \cdot 3\sqrt{5}}{5 \cdot 6} = 1 - 0 \cdot 12 = 0 \cdot 88$$

$$\therefore \quad \bar{P} = 0 \cdot 88P = 0 \cdot 88 \times 25 = 22 \text{ mm in 1 hour}$$

Check from Oxford's figures: From Fig. 2.7 the 10 year frequency for 60 min duration indicates 23 mm in 1 hour.

$$\therefore \quad \bar{P} = 0 \cdot 88 \times 23 = \text{about 20 mm in 1 hour}$$

In the assessment of total quantities of rainfall over large areas, the incidence of particular storms and their contribution to particular gauges is unknown, and it is necessary to convert many point values to give an average rainfall depth over a certain area. The simplest way of doing this is to take the arithmetical mean of the amounts known for all points in the area. If the distribution of such points over the area is uniform and the variation in the individual gauge's amounts are not large, then this method gives reasonably good results.

Another method, due to Thiessen [12], defines the zone of influence of each station by drawing lines between pairs of gauges, bisecting the lines with perpendiculars, and assuming all the area enclosed within the boundary formed by these intersecting perpendiculars has had rainfall of the same amount as the enclosed gauge.

A variation of this technique is to draw the perpendiculars to the lines joining the gauges at points of median altitude, instead of at mid length. This *altitude-corrected* analysis is sometimes held to be a more logical approach but as a rule produces little difference in result. Either method is more accurate than that of the simple arithmetic mean but involves much labour. Thiessen polygons are illustrated in Fig. 2.11.

A third method is to draw *isohyets*, or contours of equal rainfall depth. The areas between successive isohyets are measured and assigned an average value of rainfall. The overall average for the area is thus derived from weighted averages. This method is possibly the best of the three and has the advantage that the isohyets may be drawn to take account of local effects like prevailing wind and uneven topography. A typical isohyetal map is shown in Fig. 2.12 though the fall recorded is far from typical, this being the heaviest recorded daily fall in the United Kingdom.

2.8.5 Supplementing rainfall records. It frequently happens when assembling rainfall data that there are areas inadequately recorded, particularly regarding intensities of rainfall. For example, suppose that at two rainfall stations A and B, there is a recording gauge at A and a non-recording gauge at B. Suppose the mass curve of rainfall at A is as shown in the full line on the graph of Fig. 2.13. The total

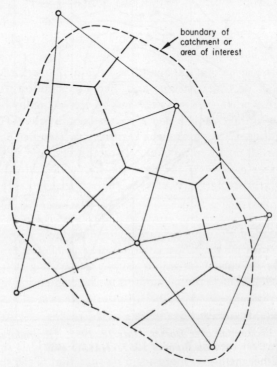

boundary of
catchment or
area of interest

The area assumed to have the rainfall of a particular
gauging station is enclosed by the dashed lines and
catchment boundary

FIG. 2.11 *Thiessen polygons*

rainfall at B is known and appears as a point on this graph. If the physical location of B is near A and its rainfall is likely to be of the same kind and frequency, then it is permissible to assume that the mass curve of B will be as shown by the dashed line on the graph. This kind of extension of data should be used with care but can be very useful.

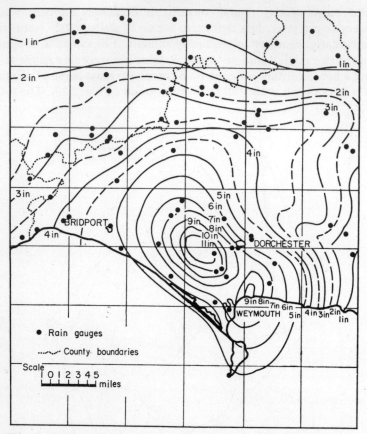

FIG. 2.12 *Rainfall over Dorset. July 18, 1955. (Reproduced from British Rainfall 1955. HMSO 1957)*

Another example of this is the filling in of a gap in a station's records, when those for neighbouring stations have provided data for the missing period. Suppose for a certain year there is no record of the precipitation at A. In the same year the total at B has been 650 mm.

Assuming that the mean annual precipitation at A and B respectively is 700 mm and 600 mm then by simple proportion, assuming the average relationship holds for the missing year also, the missing years, precipitation at A will be $\dfrac{700}{600} \times 650 = 758$ mm. This result may be checked by a similar reference to station C.

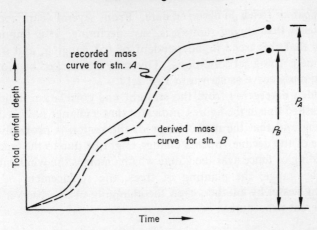

FIG. 2.13 *Derivation of rainfall data*

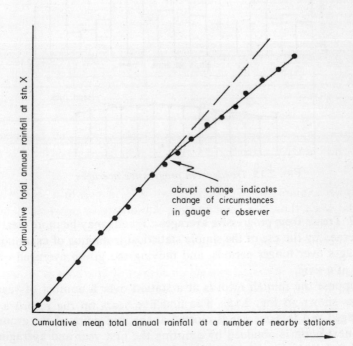

FIG. 2.14 *Station check by double mass curve*

2.8.6 Apparent trends in observed data. From several years' records it may seem that annual rainfall is, say, declining. It is important to know that this trend is independent of the gauging, and is due only to meteorological conditions. This may be checked by plotting a double mass curve as shown in Fig. 2.14.

A sudden divergence from the straight line correlation indicated by the dashed line in the figures, indicates that a change has occurred in gauging and that the meteorology of the region is probably not the cause of the decline. Such a change might be due to the erection of a building or fence near the gauge which changes the wind pattern round the gauge, the planting of trees, the replacement of one measuring vessel by another, even the changing of an observer who uses different procedures.

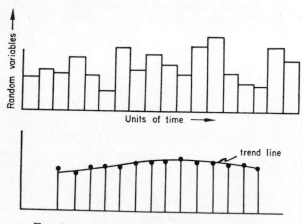

FIG. 2.15 *Trends from progressive averages*

2.8.7 Trends from progressive averages. Trends may be more clearly discerned by the use of the simple statistical technique of examining averages over longer periods, and moving the group averaged one year at a time.

Suppose the rainfall records at a station over a number of years are as shown in Fig. 2.15. The first five years on the record are averaged and this average is plotted at the mid point of the group. The next point is obtained by omitting the first year and averaging years 2–6 inclusive, again plotting the average at the mid point of the group. In this way the wide variations of particular years are smoothed out and long term trends may be detected.

The same techniques may be applied to temperature, hours of sunshine, wind speeds, cloud cover and other data.

Rainfall maps for Great Britain and Ireland are printed on the following two pages; by courtesy of the Meteorological Office and the Irish Meteorological Service respectively. Using these maps, the long-term average annual rainfall for any locality may be determined, for use with Fig. 2.6. Note that different units are used for the two maps.

AVERAGE ANNUAL RAINFALL
1916 — 1950

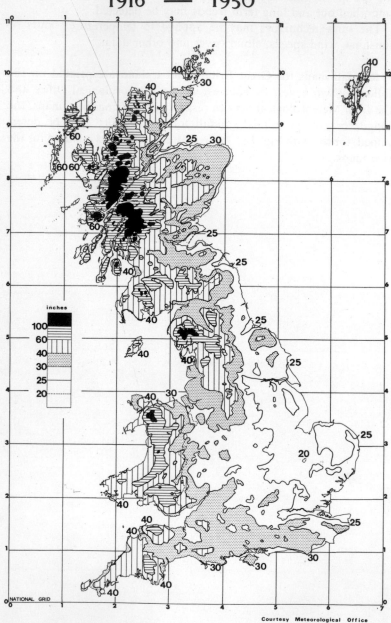

inches

100
60
40
30
25
20

NATIONAL GRID

Courtesy Meteorological Office

AVERAGE ANNUAL RAINFALL
1931 — 1960

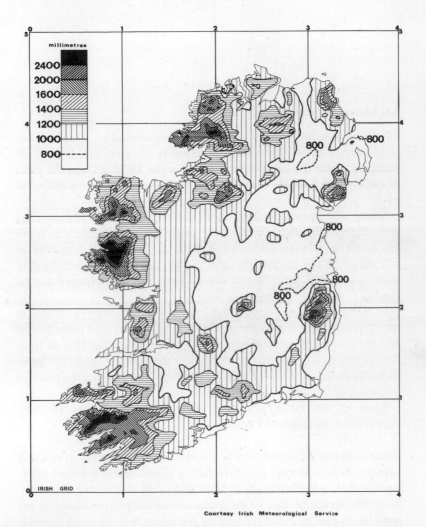

millimetres

2400
2000
1600
1400
1200
1000
800

IRISH GRID

Courtesy Irish Meteorological Service

3 Evaporation and Transpiration

3.1 Meteorological factors

Evaporation is important in all water resource studies. It affects the yield of river basins, the necessary capacity of reservoirs, the size of pumping plant, the consumptive use of water by crops and the yield of underground supplies, to name only a few of the factors affected by it.

Water will evaporate from land, either bare soil or soil covered with vegetation, also from trees, impervious surfaces like roofs and roads, open water and flowing streams. The rate of evaporation will vary with the colour and reflective properties of the surface (the *albedo*) and will be different for surfaces directly exposed to, or shaded from, solar radiation.

In moist temperate climates the loss of water through evaporation may typically be 600 mm per year from open water and perhaps 450 mm per year from land surfaces. In an arid climate, like that of Iraq, the corresponding figures could be 2000 mm and 100 mm, the great disparity in this latter case being caused by absence of precipitation for much of the year.

Some of the more important meteorological factors affecting evaporation are discussed below.

Solar radiation. Evaporation is the conversion of water into water vapour. It is a process which is taking place almost without interruption during the hours of daylight and often also during the night. Since the change of state of the molecules of water from liquid to gas requires an energy input (known as the latent heat of vaporisation), the process is most active under the direct radiation of the sun. It follows that clouds which prevent the full spectrum of the sun's radiation reaching the earth's surface will reduce the energy input and so slow up the process of evaporation.

Wind. As the water vaporises into the atmosphere the boundary layer between earth and air becomes saturated and it must be removed and continually replaced by dryer air if evaporation is to proceed. This movement of the air in the boundary layer depends on wind and so the wind speed is important.

Relative humidity. The third factor affecting evaporation is the relative humidity of the air. As the air's humidity rises, its ability to absorb more water vapour decreases and the rate of evaporation slows. Replacement of the boundary layer of saturated air by air of equally high humidity will not maintain the evaporation rate: this will only occur if the incoming air is drier than that which is displaced.

Temperature. As mentioned above, an energy input is necessary for evaporation to proceed. It follows that if the ambient temperature of the air and ground is high, evaporation will proceed more rapidly than if they are cool, since heat energy is more readily available. Since the capacity of air to absorb water vapour increases as its temperature rises, so air temperature has a double effect on how much evaporation takes place, while ground and water temperatures have single direct effects.

3.2 Transpiration

Growing vegetation of all kinds needs water to sustain life, though different species have very different needs. Only a small fraction of the water needed by a plant is retained in the plant structure. Most of it passes through the roots to the stem or trunk and is *transpired* into the atmosphere through the leafy part of the plant.

In field conditions it is practically impossible to differentiate between evaporation and transpiration if the ground is covered with vegetation. The two processes are commonly linked together and referred to as *evapo-transpiration*.

The amount of moisture which a land area loses by evapo-transpiration depends primarily on the incidence of precipitation, secondly on the climatic factors of temperature, humidity etc. and thirdly on the type, manner of cultivation and extent of vegetation. The amount may be increased, for example, by large trees whose roots penetrate deeply into the soil, bringing up and transpiring water which would otherwise be far beyond the influence of surface evaporation.

Transpiration proceeds almost entirely by day under the influence of solar radiation. At night the pores or *stomata* of plants close up and very little moisture leaves the plant surfaces. Evaporation, on the other hand, continues so long as a heat input is available, and accordingly primarily during the day. The other factor of importance is the availability of a plentiful water supply. If water is always available in abundance for the plant to use in transpiration, more will be used than if at times less is available than could be used. Accordingly, a distinction must be made between *potential evapo-transpiration* and what may actually take place. Most of the methods of estimation necessarily assume an abundant water supply and so give the potential figure.

3.3 Methods of estimating evaporation

3.3.1 Water budget or storage equation approach. This method consists of drawing up a balance sheet of all the water entering and leaving a particular catchment or drainage basin. If rainfall is measured over the whole area on a regular and systematic basis then a close approximation to the amount of water arriving from the atmosphere may be made. Regular stream gauging of the streams draining the area and accurately prepared flow-rating curves will indicate the water leaving the area by surface routes. The difference between these two can only be accounted for in three ways:

(i) by a change in the storage within the catchment, either in surface lakes and depressions or in underground aquifers;
(ii) by a difference in the underground flow into and out of the catchment;
(iii) by evaporation and transpiration.

The storage equation can be written generally as

$$E = P + I \pm U - O \pm S \qquad (3.1)$$

where E = evapo-transpiration
     P = total precipitation
     I = surface inflow (if any)
     U = underground outflow
     O = surface outflow
     S = change in storage (both surface and subsurface)

If the observations are made over a sufficiently long time the significance of S, which is not cumulative, will decrease and may be ignored, if the starting and finishing points of the study are chosen to coincide as nearly as possible with the same seasonal conditions. The significance of U cannot be generalised but in many cases can be assigned second order importance because of known geological conditions which preclude large underground flows. In such cases a good estimation of evapo-transpiration becomes possible and the method provides a means of arriving at first approximations.

3.3.2 Energy budget method. This method, like the water budget approach, involves solving an equation which lists all the sources and sinks of thermal energy and leaves evaporation as the only unknown. It involves a great deal of instrumentation and is still under active development. It cannot readily be used without much data that is not normally available and so it is a specialist approach.

3.3.3 Empirical formulae. Many attempts have been made to produce satisfactory formulae for the estimation of evaporation. These are usually for evaporation from an open water surface, as indeed, are the more general methods to follow. The reason for this is simple. Evaporation if it is to take place, presupposes a supply of water. Whatever the meteorological conditions may be, if there is no water present, then there can be no evaporation. Accordingly, estimating methods using meteorological data work on the assumption that abundant water is available, i.e. a free water surface. The results obtained therefore are not necessarily a measure of actual, but of *potential evaporation*. Often these two are the same, as for example, in reservoirs where a free water surface exists. When evaporation from land surfaces is concerned, the loss of water in this way clearly depends on availability: rainfall, water-table level, crop or vegetation, and soil type all have an influence which may be expressed by applying an empirical factor, usually less than unity, to the free water surface evaporation.

There are two cases that should be considered:

(i) when the temperature of the water surface is the same as the air temperature;

(ii) when the air and water surface temperatures are different.

Case (i) rarely occurs and is empirically treated by the equation

$$E_a = C(e_s - e)f(u) \qquad (3.2)$$

where E_a = open water evaporation per unit time (for air and water temperature the same $t°C$) in mm/day

C = an empirical constant

e_s = saturation vapour pressure of the air at $t°C$ (mm mercury)

e = actual vapour pressure in the air above (mm mercury)

u = wind speed at some standard height (m)

The following equation has been empirically obtained for this case and is of general validity

$$E_a = 0.35(e_s - e)(0.5 + 0.54u_2) \qquad (3.3)$$

u_2 denotes wind speed in m/s at a height of 2 m: E_a is in mm/day.

Case (ii) is the normally occurring case. Again a formula should have the form

$$E_o = C(e'_s - e)f(u) \qquad (3.4)$$

but now e'_s is the saturation vapour pressure of the boundary layer of air between air and water, whose temperature t'_s is not the same as either air or water and is virtually impossible to measure. Accordingly empirical formulae have been developed in the form of Eq. (3.2) which works fairly well for specific locations where the constants have been derived, but have no general validity.

Such a formula, derived for the Ijsselmeer in Holland, and *only applicable to it and similar conditions*, is

$$E_o = 0.345(e_w - e)(1 + 0.25u_6) \qquad (3.5)$$

where E_o = evaporation of the lake in mm per day

e_w = saturation vapour pressure at temperature t_w of the surface water of the lake in mm mercury

e = actual vapour pressure in mm mercury

u_6 = wind velocity in m/s at a height of 6 m above the surface.

3.3.4 Penman's Theory. The following nomenclature is used

E_o = evaporation from open water (or its equivalent in heat energy)

e_w = saturation vapour pressure of air at water surface temperature t_w

e = actual vapour pressure of air at temperature t = saturation vapour pressure at dew point t_d

e_s = saturation vapour pressure of air at temperature t

e'_s = saturation vapour pressure of air at boundary layer temperature t'_s

n/D = cloudiness ratio = actual/possible hours of sunshine

R_A = Angot's value of solar radiation arriving at the atmosphere

R_C = sun and sky radiation actually received at earth's surface on a clear day

R_I = nett amount of radiation absorbed at surface after reflection

R_B = radiation *from* the earth's surface.

In 1948 H. L. Penman presented a theory and formula [13] for the estimation of evaporation from weather data. The theory is based on two requirements which must be met if continuous evaporation is to occur. These are (i) that there must be a supply of energy to provide latent heat of vaporisation and (ii) there must be some mechanism for removing the vapour, once produced.

The energy supply. During the hours of daylight there is a certain measurable amount of short wave radiation arriving at the earth's surface. The amount depends on latitude, season of the year, time of day and degree of cloudiness. Assuming there were no clouds and a perfectly transparent atmosphere, the total radiation to be expected at a point has been given in tabular form by Angot, and is reproduced in Table 3.1 as values of R_A.

If R_C = short wave radiation actually received at the earth from sun and sky and n/D = ratio of actual/possible hours of sunshine then Penman gives

$$R_C = R_A(0.18 + 0.55n/D)$$

for southern England

TABLE 3.1. *Angot's values of short-wave radiation flux R_A at the outer limit of the atmosphere in gramme-calories per cm² per day, as a function of the month of the year and the latitude*

Latitude°	Jan.	Feb.	Mar.	Apr.	May	June	July	Aug.	Sept.	Oct.	Nov.	Dec.	Year
N 90	0	0	55	518	903	1077	944	605	136	0	0	0	3540
80	0	3	143	518	875	1060	930	600	219	17	0	0	3660
60	86	234	424	687	866	983	892	714	494	258	113	55	4850
40	358	538	663	847	930	1001	941	843	719	528	397	318	6750
20	631	795	821	914	912	947	912	887	856	740	666	599	8070
Equator	844	963	878	876	803	803	792	820	891	866	873	829	8540
20	970	1020	832	737	608	580	588	680	820	892	986	978	8070
40	998	963	686	515	358	308	333	453	648	817	994	1033	6750
60	947	802	459	240	95	50	77	187	403	648	920	1013	4850
80	981	649	181	9	0	0	0	0	113	459	917	1094	3660
S 90	995	656	92	0	0	0	0	0	30	447	932	1110	3540

From: *Physical and Dynamical Meteorology* by David Brunt, p. 112 (Cambridge University Press, 1944). The SI unit for R_A is joules/m²/day. The table in gm cal/cm²/day is used so that it is compatible with Rijkoort's nomogram. The conversion is

$$1 \text{ gm cal/cm}^2 = 41 \cdot 9 \text{ kJ/m}^2$$

40

and quotes Kimball

$$R_C = R_A(0.22 + 0.54n/D)$$

for Virginia, U.S.A.

and Prescott

$$R_C = R_A(0.25 + 0.54n/D)$$

for Canberra, Australia

from which it may be seen that even on days of complete cloud cover ($n/D = 0$), about 20% of solar radiation reaches the earth's surface, while on cloudless days about 75% of radiation gets through.

Part of R_C is reflected as short wave radiation; how much depends on the reflectivity of the surface, or, the reflection coefficient r.

If R_I = the nett amount of radiation absorbed

then $\quad R_I = R_C (1 - r) = R_A(1 - r)(0.18 + 0.55n/D)$

In turn, some of R_I is re-radiated by the earth as long wave radiation, particularly at night when the air is dry and the sky clear. The nett outward flow R_B may be expressed empirically as

$$R_B = \sigma T_a{}^4(0.47 - 0.077\sqrt{e})(0.20 + 0.80n/D)$$

where σ = Lummer and Pringsheim constant = 117.74×10^{-9} g cal/cm²/day

T_a = absolute earth temperature = $t°C + 273$

e = actual vapour pressure of air in mm mercury

and so the nett amount of energy finally remaining at a free water surface ($r = 0.06$) is given by H, where

$$\begin{aligned} H &= R_I - R_B \\ &= R_C - rR_C - R_B \\ &= R_C(1 - r) - R_B \\ &= R_A(0.18 + 0.55n/D)(1 - 0.06) - R_B \end{aligned}$$

$\therefore \ H = R_A(0.18 + 0.55n/D)(1 - 0.06) - (117.4 \times 10^{-9})$

$$T_a{}^4(0.47 - 0.077\sqrt{e})(0.20 + 0.80n/D) \qquad (3.6)$$

and this heat is used up in four ways

i.e. $\qquad\qquad H = E_o + K + S + C \qquad\qquad (3.7)$

where E_0 = heat available for evaporation from open water

        K = convective heat transfer from the surface

        S = increase in heat of the water mass (i.e. storage)

        C = increase in heat of the environment (negative advected heat)

Over a period of days and frequently over a single day, the storage of heat is small compared with the other changes, and the same is true of environmental storage, so that to a small degree of error

$$H = E_0 + K \tag{3.8}$$

Vapour removal. It has been shown that evaporation may be represented by

$$E_0 = C(e'_s - e)f(u) \tag{3.4}$$

but that e'_s cannot be evaluated if air and water are at different temperatures. Penman now made the assumption that the transport of vapour and the transport of heat by eddy diffusion are essentially controlled by the same mechanism, i.e. atmospheric turbulence, the one being governed by $(e'_s - e)$, the other by $(t'_s - t)$. To a close approximation therefore

$$\frac{K}{E_0} = \beta = \frac{\gamma(t'_s - t)}{e'_s - e} \qquad \begin{aligned} &\text{where } \gamma = \text{psychrometer constant} \\ &\quad\quad = 0 \cdot 66 \text{ if } t \text{ is } °\text{C} + e \text{ in mbar} \end{aligned}$$

Now since

$$H = E_0 + K = E_0(1 + \beta)$$

$$\therefore \qquad E_0 = \frac{H}{1 + \beta} = \frac{H}{1 + \gamma \dfrac{t'_s - t}{e'_s - e}}$$

Now eliminate $t'_s - t$ by substitution, since $t'_s - t = \dfrac{e'_s - e_s}{\Delta}$ where

e_s = saturation vapour pressure at temperature t.

Δ = slope of vapour pressure curve at t, = $\tan \alpha$ (see Fig. 3.1). This is reasonable since t'_s is never very far from t.

Hence
$$E_0 = \frac{H}{1 + \dfrac{\gamma}{\Delta} \cdot \dfrac{e'_s - e_s}{e'_s - e}} \tag{3.9}$$

Now e_s must be eliminated.

Since
$$e'_s - e_s = (e'_s - e) - (e_s - e) \tag{3.10}$$

and from Eq. 3.2 $\qquad E_a = Cf(u)(e_s - e)$

and Eq. 3.4 $\qquad E_o = Cf(u)(e'_s - e)$ $\left.\begin{array}{c} \\ \\ \end{array}\right\} \therefore \dfrac{E_a}{E_o} = \dfrac{e_s - e}{e'_s - e}$ (3.11)

where E_a = evaporation (in energy terms) for the hypothetical case of equal temperatures of air and water,

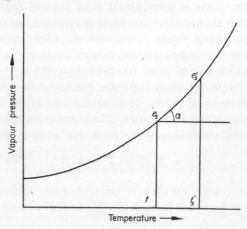

FIG. 3.1 *Saturation vapour pressure curve*

then by substituting in Eq. (3.9) the values of Eqs. (3.10) and (3.11)

$$E_o = \frac{H}{1 + \dfrac{\gamma}{\Delta}\left[\dfrac{(e'_s - e) - (e_s - e)}{e'_s - e}\right]}$$

and

$$E_o = \frac{H}{1 + \dfrac{\gamma}{\Delta}\left(1 - \dfrac{E_a}{E_o}\right)}$$

from which $\qquad E_o = \dfrac{\Delta H + \gamma E_a}{\Delta + \gamma}$ (3.12)

Δ has values obtained from the saturation vapour pressure curve, typically as shown

$t =$		$\Delta =$
0°C		0·36
10		0·61
20		1·07
30		1·80

Referring to Eqs. (3.3) and (3.6) for E_a and H respectively, it may be seen that E_o is now computed from standard meteorological observations of mean air temperature, relative humidity, wind velocity at a standard height, and hours of sunshine. The formula has been checked in many parts of the world and gives very good results. Being based on physical principles it is of general application and gives values that should serve for most project studies until supplemented by actual evaporation measurements. (See Section 3.6).

To overcome the computational labour involved in solving the Penman equation, a nomogram has been designed by P. J. Rijkoort of the Royal Meteorological Institute, Netherlands, which enables rapid evaluations to be made. It is reproduced at the back of this book by permission of the designer. The nomogram has been drawn for a slightly different value of R_C from that used by Penman

$$R_C = R_A(0 \cdot 20 + 0 \cdot 48 n/D)$$

instead of Penman's $R_A(0 \cdot 18 + 0 \cdot 55 n/D)$ but any difference will be smaller than the probable margin of error in cloud cover estimation, so it may be ignored.

Values of R_A may be derived for any latitude from Table 3.1, p. 40.

3.4 Evaporation from land surfaces using Penman's E_o value

Penman in his original paper reported on results of experiments conducted on turfed soil and bare soil to determine how their evaporation rates (E_T and E_B) compared with open water (E_o). He concluded that the evaporation rate from a freshly wetted bare soil was about 90% of that from an open water surface exposed to the same weather.

i.e. $E_B/E_o = 0 \cdot 90$

For grassed surfaces the comparison was more erratic, and provisional figures for turf with a plentiful water supply were given as follows:

Values of E_T/E_o for southern England

Nov.–Feb.	0·6
Mar.–Apr. ⎫ Sept.–Oct. ⎬	0·7
May–Aug.	0·8
Whole year	0·75

These figures are all less than unity because of the greater reflectivity of vegetation compared to open water and also because the transpiration of plants virtually ceases at night.

3.5 Thornthwaite's formulae for evapo-transpiration

C. W. Thornthwaite carried out many experiments in the United States using lysimeters and studied extensively the correlation between temperature and evapo-transpiration. From this work [14] he devised a method enabling estimates to be made of the potential evapo-transpiration from short, close set vegetation with an adequate water supply, in the latitudes of the United States.

If t_n = average monthly temperature of the consecutive months of the year in °C (where $n = 1, 2, 3, \ldots 12$) and j = monthly 'heat index',

then
$$j = \left(\frac{t_n}{5}\right)^{1\cdot514} \tag{3.13}$$

and the yearly 'heat index', J is given by

$$J = \sum_{1}^{12} j \quad \text{(for the 12 months)} \tag{3.14}$$

The potential evapo-transpiration for any month with average temperature $t°$ is then given, as PE_x, by

$$PE_x = 16 \left(\frac{10t}{J}\right)^a \text{ mm per month} \tag{3.15}$$

where

$$a = (675 \times 10^{-9})J^3 - (771 \times 10^{-7})J^2 + (179 \times 10^{-4})J + 0\cdot492 \tag{3.16}$$

However PE_x is a theoretical standard monthly value based on 30 days and 12 h sunshine per day. The actual PE for the particular month with average temperature $t°$ is given by

$$PE = PE_x \frac{DT}{360} \text{ mm} \tag{3.17}$$

where D = number of days in the month
T = average number of hours between sunrise and sunset in the month.

The method has been tested by Serra, who suggested that Eqs. (3.13) and (3.16) may be simplified as follows

$$j = 0 \cdot 09 t_n^{3/2} \tag{3.18}$$

$$a = 0 \cdot 016 J + 0 \cdot 5 \tag{3.19}$$

This method of estimating potential evapo-transpiration is empirical and complicated and requires the use of a nomogram for

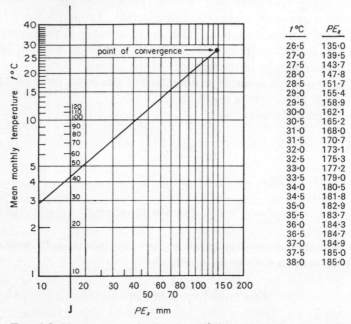

FIG. 3.2 *Nomogram and table for finding potential evapo-transpiration PE_x (after C. W. Thorn·hwaite. Courtesy of "Geog. Review"; copyright American Geographical Society, New York)*

its solution. Thornthwaite published such a nomogram which is reproduced in Fig. 3.2.

The first step is to obtain the heat index J. From Fig. 3.2 obtain the unadjusted value of potential evapo-transpiration by drawing a straight line from the location's J value through the point of convergence at $t = 26 \cdot 5°C$. (If $t°$ is greater than $26 \cdot 5°C$ use the Table alongside Fig. 3.2.) PE_x for the month can then be read off, corresponding to its given mean temperature. Twelve values are obtained for each of the 12 months. These unadjusted values may then be adjusted for day and month length by Eq. (3.17) and totalled to give annual potential evapo-transpiration.

It has been found that the method gives reasonably good results whatever the vegetation cover, though different types of vegetation will affect a particular locality's true value. The formula is based on temperature, which does not necessarily correspond to incoming solar radiation immediately, because of the "heat inertia" of land and water. Transpiration, however, responds directly to solar radiation. Accordingly, care should be exercised when using the method to ensure that conditions do not change abruptly in a particular month, though if figures for many consecutive months are being used, the cumulative differences are probably negligible.

For project studies the method is a useful complement to the Penman approach.

3.6 Direct measurement of evaporation by pans

Whenever possible, direct observations of evaporation should be made. The instrument used for this is the evaporation pan. In Britain the standard pan is 1·83 m (6 ft) square and 610 mm (2 ft) deep filled to a depth of 550 mm (1 ft 9 in.) and set in the ground so that the rim of the pan projects 76 mm (3 in.) above the surrounding ground. Regular observations of pan evaporation are made at some 30 points throughout the country [15].

In the U.S.A. the standard or Class A pan is circular 1·22 m (4 ft) in diameter and 254 mm (10 in.) deep, filled to a depth of 180 mm (7 in.), set on a timber grillage with the pan bottom 150 mm (6 in.) above ground level. Regular evaporation readings are taken at over 400 places.

A third type of pan is sometimes used in the United Kingdom, the Peirera pan, which is circular like the Class A pan but deeper and sunk in the ground with a 3 in. air space surrounding it.

The Class A pan has a greater daily range of temperature than the square one, but is usually homogeneous whereas the water in the square pan may stratify. Doubling the wind run may increase evaporation by up to 20%.

The relatively small capacities and shallow depths of pans in comparison to lake and river volumes and their situation at or near the land surface allows proportionately greater amounts of advected heat from the atmosphere to be absorbed by the water in the pan through the sides and bottom, than by natural open water, and by some pans more than others. Pan evaporation is therefore usually too high and a pan coefficient has to be applied. These coefficients range from 0·65 to greater than unity, depending on the dimensions and siting of the pan. Generally the standard British pan has a co-efficient about 0·92 and the U.S. Weather Bureau Class A pan about 0·75 but there are quite wide variations. Law [16] carried out comparative tests over 14 years at two sites in Yorkshire and found the ratio of evaporation from the Class A pan to that from the square British pan ranged between 1·17 and 1·40 with an average of 1·32. Houk [17] gives a full account of known American values and Olivier [18] quotes many data from African and Near Eastern sources.

There are difficulties in using pans for the direct measurement of evaporation, arising from the difficulty of measuring very small differences of elevation and the subsequent application of coefficients to relate the measurements from a small tank to large bodies of open water. Nevertheless, actual field measurements should form an important part of any project studies of evaporation.

3.7 Consumptive use

Evapo-transpiration is the term used for the evaporation of moisture from the earth's surface including lakes and streams and the vegetation that may cover the land. *Consumptive use* refers to the evaporation and transpiration from vegetation-covered land areas only, frequently in respect to horticulture and agriculture and associated irrigation requirements. The terms are often used synonymously.

The consumptive use of water in an area is dependent on many factors, including climate, the supply of soil moisture, growing vegetation, type of soil and methods of land management. Climatic factors include precipitation, temperature, humidity, wind and latitude of the locality (which affects the length of the growing season). Soil moisture supply depends on topography and underground flow, as well as precipitation. Soil types and land management vary widely over short distances. There are no formulae of general validity but several empirical formulae may be used with local coefficients to determine annual water use in any locality within certain broad limitations.

3.7.1 Arable crops. Consumptive use refers to water that is actually used while evapo-transpiration formulae give potential water use. Reference to Fig. 3.3 will show that for a particular locality, unless the rain falls when it is needed, a large water deficit may develop in the growing season despite quite high rainfall. In a case like this consumptive use will be less than potential evapo-transpiration and a need for irrigation in the growing period is indicated.

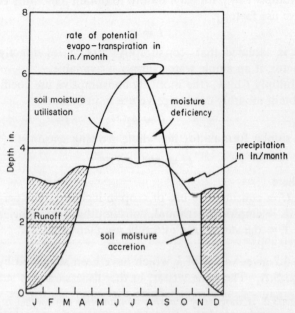

FIG. 3.3 *Typical annual soil moisture deficiency diagram*
(*after Thornthwaite* [19])

Actual measurements of consumptive use are very difficult and slow to make and may be very expensive. Many such investigations, have been made in the U.S.A., and the results may be applied in other areas by a method developed by the Division of Irrigation and Water Conservation, Soil Conservation Service of the U.S. Dept. of Agriculture [20]. The existing consumptive use data is correlated with monthly temperature, percentage of daylight hours, length of growing period and precipitation. Coefficients are developed for different crops and these coefficients are then used for crops in areas where climatological data only, are available. Although other factors are involved, the four mentioned have the principal influence and are generally available from standard meteorological observations. The procedure is as follows,

(1) Tabulate the mean monthly temperature for each month in °F (t).
(2) Compute the monthly proportion of daylight hours of the year (p).
(3) Multiply t by p for each month to obtain a monthly consumptive use factor f.

$$f = tp$$

It is assumed that, consumptive use varies directly as this factor, if an ample water supply is available.

(4) Multiply f by k, (the monthly consumptive use coefficient) to obtain monthly consumptive use u, in inches.

$$u = kf$$

A similar formula for the whole growing season is

$$U = KF = \Sigma kf$$

where
U = consumptive use of crop in inches for the period
K = empirical, seasonal, consumptive-use coefficient
F = the sum of the monthly use factors Σf

Table 3.2 gives values of k which have been suggested by Blaney for the U.S.A. The same writer, in the discussion to a subsequent

TABLE 3.2. *Examples of monthly consumptive use coefficients* (k) *for irrigated crops based on field measurements of evapo-transpiration and temperature*

Location	Crop	Mar.	April	May	June	July	Aug.	Sept.	Oct.	Nov.
Arizona	Alfalfa	0·74	0·84	0·91	1·10	1·30	—	0·90	0·75	0·75
	Cotton	—	0·30	0·40	0·60	0·80	0·80	0·70	0·60	—
	Soya beans	—	—	—	0·35	0·60	0·90	0·80	0·50	—
	Guar	—	—	—	—	0·30	0·80	0·90	0·55	—
	Grapefruit	0·55	0·65	0·65	0·70	0·70	0·75	0·70	0·70	0·65
	Oranges	0·53	0·56	0·56	0·58	0·58	0·61	0·61	0·61	0·60
California	Alfalfa	—	0·70	0·80	0·80	0·90	0·90	0·90	0·80	0·70
	Lemons	—	0·40	0·40	0·50	0·50	0·55	0·60	0·50	0·40
	Oranges	—	0·50	0·50	0·55	0·60	0·60	0·60	0·60	0·50
	Beets	—	0·30	0·60	0·85	0·95	0·90	0·40	—	—
	Tomatoes	—	—	—	0·45	0·80	0·70	0·80	0·70	—

from Harry F. Blaney—*Monthly consumptive use of water by irrigated crops and natural vegetation*. 11th General Assembly I.A.S.H., Toronto, 1957.

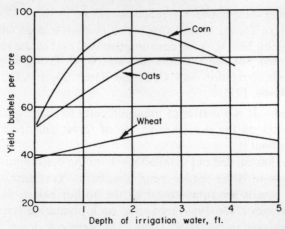

FIG. 3.4 *Water-yield relationship for crops; Cache Valley, Utah, U.S.A. (after Harris)*

paper [21] suggested that for humid areas the seasonal coefficient K should be reduced by about 10%.

Houk [17] points out that there is an optimum water requirement for every crop and that the supply of water beyond the optimum amount reduces yields. Only general guide lines can be laid down, however, because differences in soil texture and fertility affect the optimum amount for the same crop in different conditions. A typical water yield relationship for Utah, U.S.A. by Harris [22] is shown in Fig. 3.4 and the range of variation in water requirements for different crops is illustrated in a particular case by Table 3.3 which is reproduced from a report by Fortier [23].

TABLE 3.3 *Variations in water requirements of different crops produced in the Missouri and Arkansas River basins (after Fortier)*

Crop	No. of tests	Range in water requirements mm	Crop	No. of tests	Range in water requirements mm
Forage, inc. alfafa	648	590–800	Apples	4	640–795
Barley	335	405–555	Beans	4	395–490
Oats	409	410–550	Buckwheat	3	320–395
Wheat	542	415–550	Cantaloupes	10	455–705
Corn	70	375–560	Peas	168	415–590
Kafir corn	15	435–480	Potatoes	350	420–520
Flax	50	450–565	Sugar beets	128	490–765
Millet	14	245–285	Sunflowers	16	365–425
Milo maize	27	335–520	Tomatoes	6	640–855
Sorghum	26	325–450	Cucumbers	7	530–1140

For further information reference should be made to U.S. Dept. of Agriculture Technical Bulletins which contain a large amount of information on field tests for consumptive use, and to the references [17, 18] quoted. See also J. Doorenbos and W. O. Pruitt, *Crop Water Requirements*, Irrigation and Drainage Paper 24, F.A.O., United Nations, Rome, 1975.

3.7.2 Forest. It is a matter of great difficulty to make quantitative measurements on the consumptive use of forest and there is little published about it.

Law [16] has carried out detailed work on the consumptive use of a plantation of Sitka spruce near Slaidburn, Yorkshire, over 15 years. His results are summarised by the author below, as 14-year annual averages (1956–1969) measured on a plantation lysimeter of 450 m² area containing 74–95 trees. (21 were cut down half-way through the period.)

	Annual values	
	mm	%
(1) Mean rainfall from 3 rain gauges outside the plantation	1492	
(2) Estimated mean rainfall over lysimeter (based on 3 tree-top gauges)	1363	100
(3) Throughfall	923	68
(4) Interception (2) − (3)	440	32
(5) Lysimeter runoff	535	39
(6) Gross losses (2) − (5)	828	61
(7) Net losses without interception (3) − (5) or (6) − (4), i.e. transpiration	388	29

and as measured concurrently in a neighbouring climatological station

(8) Evaporation from 1·83 m square tank (mean of two neighbouring tanks)	507	
(9) Losses from adjacent grass-covered plot (measured by percolation gauge)	410	
(10) Calculated evaporation by Penman equation (mean of two calculations for two sites)	511	

From his experiments Law drew certain general conclusions

(a) Trees transpire roughly the same quantities as grass.

(b) A third of rainfall is intercepted by coniferous trees and subsequently lost through evaporation in addition to transpiration.

(c) Up to 3 mm rain can be intercepted by conifers at any one time.   (There are indications that deciduous trees do not differ much.)

(d) The evaporation of intercepted rain in coniferous forests does not differ markedly from summer to winter because the effect of wind is more important than ambient temperature, when adequate evaporating surface area is available.

(e) Stemflow on Sitka spruce is about 5% of throughfall.

Currently (1973) in the United Kingdom extensive experiments are under way by the Institute of Hydrology in the Plynlimon catchments which should add considerably to existing knowledge on the consumptive use of forest.

4 Infiltration and Percolation

4.1 Infiltration capacity of soil

When rain falls upon the ground it first of all wets the vegetation or the bare soil. When the surface cover is completely wet, subsequent rain must either penetrate the surface layers, if the surface is permeable, or run off the surface towards a stream channel if it is impermeable.

If the surface layers are porous and have minute passages available for the passage of water droplets, the water *infiltrates* into the subsurface soil. Soil with vegetation growing on it is always permeable to some degree. Once infiltrating water has passed through the surface layers, it *percolates* downward under the influence of gravity until it reaches the zone of saturation at the *phreatic surface*.

Different types of soil allow water to infiltrate at different rates. Each soil type has a different infiltration capacity, f, measured in mm/h or in./h. For example it can be imagined that rain falling on a gravelly or sandy soil will rapidly infiltrate and, provided the phreatic surface is below the ground surface, even heavy rain will not produce surface runoff. Similarly a clayey soil will resist infiltration and the surface will become covered with water even in light rains. The rainfall rate, i, also obviously affects how much rain will infiltrate and how much will run off.

4.2 Factors influencing f_c

Nassif and Wilson [24] have recently carried out extensive studies on infiltration using a weighable laboratory catchment 25 m^2 in area. They conclude that for any soil under constant rainfall, infiltration

54

rate decreases in accordance with an equation of the form first used by Horton [25].

$$f = f_c + \mu e^{-Kt}$$

where f = infiltration rate at any time t (mm/h)
f_c = infiltration capacity at large value of t (mm/h)
$\mu = f_0 - f_c$
f_0 = initial infiltration capacity at $t = 0$ (mm/h)
t = time from beginning of rainfall (min)
K = constant for a particular soil and surface (min^{-1})

K is a function of surface texture: e.g., if vegetation is present K is small, while a smoother surface texture, such as bare soil, will yield larger values.

f_0 and f_c are functions of both soil type and cover. For example, a bare sandy or gravelly soil will have high values of f_0 and f_c and a bare clayey soil will have low values of f_0 and f_c but both values will increase for both soils if they are turfed.

f_c is a function of slope up to a limiting value of slope (varying between 16 and 24%) after which there is little variation.

f_c is a function of initial moisture content: the dryer the soil initially, the larger will be f_c, but the variation may be comparatively small.

f_c is a function of rainfall intensity. If the intensity i increases, f_c increases. This parameter has a greater effect on f_c than any other variable.

These correlations are illustrated in Fig. 4.1 for a typical agricultural soil. Table 4.1 lists some representative values of K, f_0 and f_c for different soil types. The parameters K and f_0 are relatively stable for particular soils and do not vary noticeably with slope of catchment or rain intensity; f_c on the other hand, varies widely with both and so is shown as a range of values.

Until recently it had been generally thought that f_c was a constant for a particular soil but this does not seem to be so. The infiltration rate appears to be largely controlled by the surface pores. Even quite a small increase in the hydrostatic head over these pores results in an increase in the flow through the soil surface. If the surface layer is imagined as shown in Fig. 4.2 where the surface soil grains are

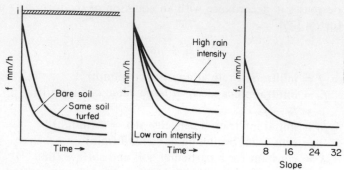

FIG. 4.1 *The variation of infiltration capacity*

shown, then the governing factor is the head h over the smallest cross section of a pore. This continues to increase with rainfall intensity until a limiting value is reached where run-off prevents any further increase. It seems unlikely that this limiting condition is often reached in natural conditions.

Previous researchers [26] have found similar results but attributed the increase in f_c at higher rainfall intensities to lack of homogeneity in their experimental catchment watershed. Others have also emphasised the over-riding importance of the *superficial layer* of a soil [27].

The infiltration rate of a soil is the sum of percolation and water entering storage above the groundwater table. Generally the soil is far from saturated and so storage goes on increasing for very long periods. Accordingly, f_c goes on decreasing under a steady rain intensity for equally long periods.

TABLE 4.1. *Representative values of K, f_0 and f_c for different soil types*

Soil type		f_0 mm/h	f_c mm/h	K min^{-1}
standard agricultural	bare	280	6–220	1·6
	turfed	900	20–290	0·8
peat		325	2–20	1·8
fine sandy clay	bare	210	2–25	2·0
	turfed	670	10–30	1·4

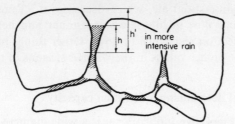

FIG. 4.2 *Hydraulic head on soil-grain pores*

It follows that arbitrary methods, described in the following sections, which for convenience, attribute average values of rain-loss to catchments to determine nett rain, are gross simplifications, at least during the early rain.

Exposed soils can be rendered almost impermeable by the compacting impact of large drops coupled with the tendency to wash very fine particles into the voids. The surface tends to become 'puddled' and the f_c value drops sharply. Similarly, compaction due to man or animals treading the surface, or to vehicular traffic, can severely reduce infiltration capacity.

Dense vegetal cover such as grass or forest tends to promote high values of f_c. The dense root systems, all providing ingress to the subsoil, the layer of organic debris forming a sponge-like surface, burrowing animals and insects opening up ways into the soil, the cover preventing compaction and the vegetation's transpiration removing soil moisture, all tend to help the infiltration process.

Other effects which marginally affect the issue are frost heave, leaching out of soluble salts, drying cracks which increase f_0, and

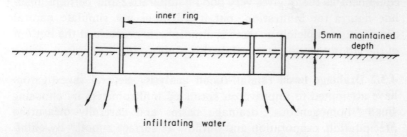

FIG. 4.3 *Infiltrometer*

entrapping of interstitial air, which decreases it. Temperature has some effect since flow in interstices is laminar and hence viscosity has a direct effect on resistance to flow. Other things being equal f_0 and f_c will have higher values in the warmer seasons of the year.

4.3 Methods of determining infiltration capacity

4.3.1 Infiltrometers. An infiltrometer is a wide diameter, short tube, or other impervious boundary surrounding an area of soil. Usually two such rings are used concentrically as shown in Fig. 4.3. The rings are flooded to a depth of 5 mm over the surface and continually refilled to maintain this depth, the inflow to the central tube being measured. The purpose of the outer tube is to eliminate to some extent the edge effect of the surrounding dryer soil. Such tests give useful comparative results but they do not simulate real conditions and have been largely replaced by sprinkler tests on larger areas. Here the sprinkler simulates rainfall, and runoff from the plot is collected and measured as well as inflow, the difference being assumed to have infiltrated.

While rain-simulating sprinklers are a good deal more realistic than flooded rings, there are limitations to the reliability of results thus obtained, which usually give higher values of f than natural conditions do. For qualitative effects, e.g. comparisons between different conditions of vegetation, soil types etc. the methods are simple and effective.

Consistent and repeatable results may be obtained by using laboratory catchments with rainfall simulators, where the quantity and thickness of soil is adequately representative of nature. Nassif and Wilson [24] used 7 tonnes of soil in a layer 200 mm thick, and measured all inputs and outputs and changes of storage. Such equipment as theirs gives very good comparative and perhaps absolute figures for infiltration, but it still does not simulate natural conditions completely as there is atmospheric pressure at the bottom of the laboratory soil layer while this is not so in nature.

4.3.2 Drainage basin rainfall-runoff analysis. Several investigators have attempted to improve on sprinkled infiltrometers by choosing small "homogeneous" drainage basins and carefully measuring precipitation, evaporation and outflow as surface runoff. By eliminating everything except infiltration, average f values may be obtained for such basins by techniques presented by Horton [28] and Sherman [29].

The difficulty remains of ensuring that there has been no unrecorded underground outflow, nor variation in underground aquifer storage, so that although quantitative results are obtained, the analysis is intricate and the margin for error is wide.

4.3.3 Φ-index method. In practice, it is possible to obtain *infiltration indices* which enable reasonable approximations to be made of infiltration losses. One of these is the Φ-*index*, which is defined as the average rainfall intensity above which the volume of rainfall equals the volume of runoff. In Fig. 4.4 a rainstorm is shown plotted

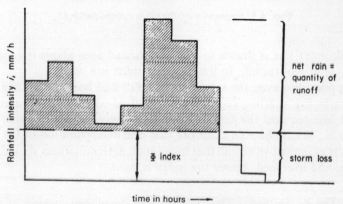

FIG. 4.4 *Infiltration loss by Φ-index*

on a time base in terms of its average hourly intensity. The shaded area above the dashed line represents measured runoff, as mm, over the catchment area. Since the unshaded area below the line is measured rainfall but did not appear as runoff, it represents all the losses, including surface detention and evaporation as well as infiltration. However, infiltration is much the largest loss in many catchments and although it is a rough and ready method (since it takes no account of the variation in f with time) it is widely used as a means of quickly assessing probable run-off from large catchments for particular rainstorms.

Example 4.1.

Given a total rainfall of 75 mm as shown in Fig. 4.5(a) and a surface runoff equivalent to 33 mm, establish the Φ-index for the catchment.

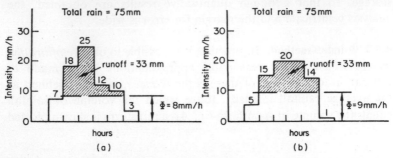

FIG. 4.5 *Examples of Φ-index computations*

The Φ-index line is drawn so that the shaded area above it contributes 33 mm of runoff. In this case the index is 8 mm/h.

Suppose, however, the same total rainfall had been distributed as shown in Fig. 4.5(b). To obtain a runoff of 33 mm above the Φ-index line requires the line to be raised to give an index value of 9 mm/h. It may be seen therefore that one determination of the Φ-index is of limited value and that many such determinations should be made, and averaged, before the index is used.

4.3.4 The f_{av} method. This method is a developed version of the Φ-index, in that it attempts to allow for depression storage and short rainless periods during a storm, as well as eliminating all rain periods where the rainfall intensity is less than the infiltration capacity assumed.

Referring to Fig. 4.6 the approximate position of the f_{av} line is fixed by reference to runoff data and the raingraph. The line is then moved vertically until the various losses are balanced and the runoff values satisfied. The loss estimates are based on whatever data is available and on the judgment of the analyst and are obviously subjective but, from the analysis of many storms, values of f_{av} may be deduced for various conditions.

In applying a derived value of f_{av} to a rainstorm to predict expected runoff, the rain periods lying outside the f_{av} period are assumed lost and the nett rain is found directly after inserting the estimated losses as shown. Butler [30] gives a detailed account of the method.

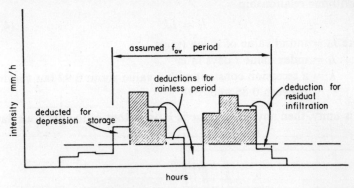

FIG. 4.6 *Example of* f_{av} *distribution*

4.4 The effect of soil moisture

4.4.1 Initial soil-moisture conditions. The foregoing methods of estimating losses are essentially based on records of rainfall and runoff for particular catchments and their behaviour under rains of varying intensity. They show average limiting values of infiltration capacity, obtaining this information in terms of the whole catchment, rather than by sampling very small areas as the infiltrometer does. They do not, however, enable predictions to be made with any accuracy of the amount of rainfall that will be absorbed by the soil and so lost to runoff in a particular case, since this depends *inter alia* on the state of wetness of the soil at the start of rainfall. As well as affecting runoff through storage capacity, the initial soil-moisture affects the infiltration capacity and hence the runoff in the initial stages of a storm. Some other measurement of this parameter is therefore necessary, if forecasts of runoff are to be made from stipulated rainfalls. Φ-index or f_{av} methods will only provide average values which may, in a particular case, be far removed from actuality, and these methods are best used after a separate assessment of initial losses.

There are two approaches to this problem discussed here. The first is the *Antecedent precipitation index*, used in the U.S.A. and the second, the *Estimated soil moisture deficit* used in Britain.

4.4.2 Antecedent precipitation index.

The antecedent precipitation index is based on the premise that soil moisture is depleted at a rate proportional to the amount in storage in the soil. There is therefore a logarithmic relationship

$$I_t = I_0 k^t \tag{4.1}$$

where I_0 = initial value of index (mm)

I_t = index value t days later

k = a recession constant with a value about 0·92 but varying between 0·85 and 0·98

If t is unity, then any day's value is k times that of the previous day.

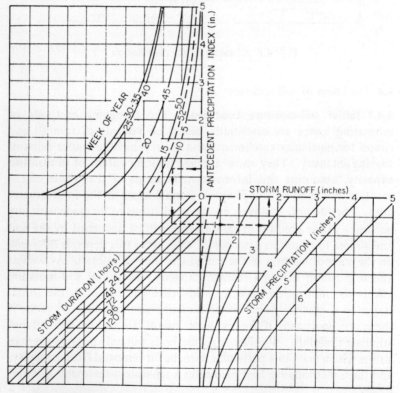

FIG. 4.7 *Storm-runoff relationship for Monacacy River at Jug Bridge, Md., U.S.A. (after U.S. Weather Bureau. From "Hydrology for Engineers" by Linsley, Kohler and Paulhus. McGraw-Hill Book Co., 1958)*

If precipitation occurs it will increase the value of the index by an amount which is indeterminate since some rain may have left the catchment as surface runoff. The amount added to the index should therefore strictly be the *basin recharge* only, but the difference in the index by using all the precipitation is usually small.

The progressive daily reduction of the index is due to evapo-transpiration, which alters seasonally, so Eq. (4.1) is used with a k value which also varies seasonally. When the index is used to assess the runoff which takes place from a particular rainstorm, this variation may be incorporated into a graphical coaxial relationship derived from the analysis of a large number of observed rainfall and runoff data on a particular catchment. Linsley, Kohler and Paulhus [31] give a detailed description of how such relationships may be derived, and Fig. 4.7 is reproduced above to illustrate the technique.

The diagram is entered at the antecedent precipitation index, and a horizontal line is followed until the particular week-number curve corresponding to the calendar date is met. From the intersection a line drops vertically down to intersect with the line of appropriate storm duration in hours, and then horizontally across to intersect with the line of total storm precipitation. A vertical line from the last intersection indicates the appropriate runoff.

The antecedent precipitation index is a valuable tool in obtaining probable runoff forecasts for well documented catchments but much work is necessary to derive relationships like Fig. 4.7.

4.4.3 Estimated soil-moisture deficit. Since evapo-transpiration is continually removing the soil-moisture and precipitation replacing it, continuous regular measurements of these two processes yield a means of estimating the *soil-moisture deficit* (or s.m.d.) without the need for an assumption such as Eq. (4.1).

When evaporation exceeds precipitation, vegetation draws on accumulated soil-moisture to continue transpiration and as the s.m.d. increases so it becomes increasingly difficult for vegetation to transpire. Different plants and crops with different root systems continue to transpire at their potential rates for different periods in the same environment. In the s.m.d. estimation carried out twice a month by the U.K. Meteorological Office [32], it is assumed that each station where rainfall is measured and evaporation estimated is representative of an area, 50% of which is covered by short-rooted

vegetation which can draw up to 3 inches moisture before actual
evapo-transpiration begins to fall below potential, 30% is covered
by long-rooted vegetation which can similarly draw up to 8 inches
moisture and 20% is riparian where the phreatic surface is so near

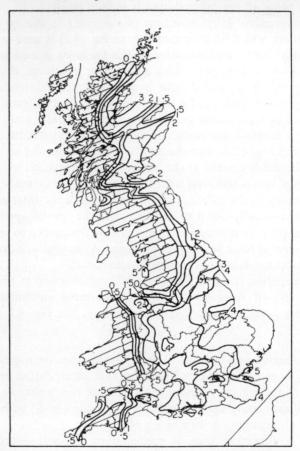

FIG. 4.8 *Estimated soil-moisture deficit in inches at 0090 GMT*
27th Sept. 1967 over Great Britain (Meteorological Office:
Crown copyright)

the surface that evapo-transpiration is never inhibited. When the
difference between rainfall and evapo-transpiration reaches 3 inches
the subsequent calculated evaporation figures take account of the
different rates and areas and a weighted figure for the whole catchment is computed.

Estimates of s.m.d. are made from observed rainfall and estimated evaporation at a network of 176 stations and a map showing Estimated Soil Moisture Deficit over the whole country is published within a day of the values at each station being received. Fig. 4.8 is a typical example. Further details of the theories used in the estimations, and methods used in compiling the data are available [33, 34].

Reference to Fig. 4.8 will show that the likelihood of a flood condition (for example) arising in any of the hydrometric areas (shown in fine subdivisions of the land area) may be gauged with much greater certainty in the knowledge of the s.m.d. there. The estimated s.m.d. is kept up to date by daily, or weekly, additions and subtractions of precipitation and evapo-transpiration, until the next issue of the published map.

The use of estimated s.m.d. to predict the proportion of runoff arising from particular storms does not differ from that for antecedent precipitation index, except in so far as coaxial relationships do not require a week number or calendar date quadrant. Actual catchment discharges are calculated in the ways described later in Sections 7.7 and 7.8, but the losses to be deducted from actual precipitation, to give nett rain, are obtained by use of estimated s.m.d. and f_{av} (or Φ-index) infiltration loss data.

4.4.4 Measurement of soil moisture. An instrument for *in situ* measurement of soil moisture, the Wallingford soil-moisture probe, has been developed by the U.K. Institute of Hydrology in conjunction with the Atomic Energy Authority. The instrument is designed for use in the field in all kinds of terrain and in any weather.

The instrument consists of a radioactive source in a 740 mm long probe which can be lowered into an aluminium access tube permanently installed in the ground, a shield and housing for the probe, a suspension cable and meter. Fast neutrons emitted from the radioactive source are scattered and slowed by collisions with the atomic nuclei of soil constituents, mainly by the hydrogen of water in the soil, thus producing "slow" neutrons. These are sensed by a slow-neutron detector in the probe and converted into electrical pulses which pass through the suspension cable to a meter where a visual display indicates the rate of detection. The wetter the soil the larger the number of collisions and hence of slow-neutrons detected. The displayed detection rate is therefore a function of the soil moisture at the probe depth. The moisture value indicated represents the mean value for a somewhat indefinite "sphere of influence" within the soil surrounding the source, with a radius which may be regarded as being about 150–300 mm. A series of readings is taken down the profile at intervals usually of 100 mm or 150 mm. The weighted

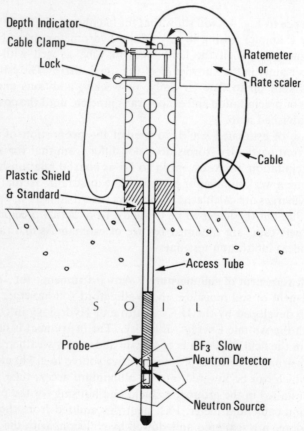

FIG 4.9 *Wallingford soil moisture probe (Courtesy Inst. of Hydrology)*

mean of these readings gives the total moisture storage in a given section of the profile.

If a catchment is provided with a number of permanently installed aluminium tubes at suitable locations, then a single battery-powered instrument can be carried from site to site to measure the *in situ* soil

FIG. 4.10 *Wallingford soil-moisture probe*

moisture at each one, thus removing much of the guess-work from the evaluation of this parameter.

The neutron probe is normally used for making repeated measurements of moisture *change* at the same site and depth and for such a purpose can give a very high precision. The accuracy of absolute moisture values on the other hand is heavily dependent upon accurate calibration at each site and depth and this is generally impracticable.

A diagram of the instrument and a photograph of one in use are shown in Figs. 4.9 and 4.10 respectively.

5 Groundwater

5.1 The occurrence of groundwater

Rainfall which infiltrates the soil and penetrates to the underlying strata is called *groundwater*. The quantity of water which can be accommodated under the surface depends on the porosity of the sub-surface strata. The water bearing strata, called *aquifers*, may consist of unconsolidated materials like sands, gravels and glacial drift or consolidated material like sandstones and limestones. Limestone is relatively impervious but is soluble in water and so frequently has wide joints and solution passages which make the rock, *en masse*, similar to a porous rock in its capacity to hold water and act as an aquifer.

The water in the pores of an aquifer is subject to gravitational forces and so tends to flow downwards through the pores of the material. The resistance to this underground flow varies widely and the *permeability* of the material is a measure of this resistance. Aquifers with large pores such as coarse gravels are said to have a high permeability, and those with very small pores such as clay, where the pores are microscopic, have a low permeability.

As the groundwater percolates down the aquifer becomes saturated. The surface of saturation is referred to as the groundwater table or the *phreatic surface*. This surface may slope steeply and its stability is dependent on supply from above. It falls during dry spells and rises in rainy weather. The water in the aquifers is usually moving slowly towards the nearest free water surface such as a lake or river, or the sea. However, if there is an impermeable layer underlying an aquifer and this layer outcrops on the surface, then the groundwater will appear on the surface in a seepage zone or as a *spring*. It is equally possible for a groundwater aquifer to become overlain by impermeable material and so be under pressure. Such an aquifer, fed from a distance, is called a *confined aquifer* and the surface to which the water would rise if it could, is called the *piezometric surface*. Another name, used for wells drilled into such confined aquifers is *artesian wells*, and the word *artesian* is sometimes also applied to the aquifers. If the piezometric surface is above

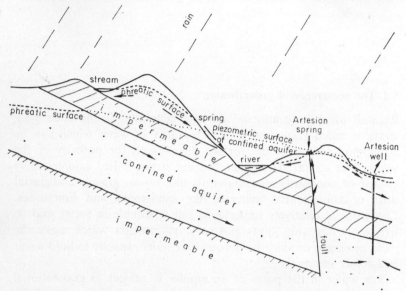

FIG. 5.1 *Modes of occurrence of groundwater*

ground level at an artesian well, it is called a flowing well, and a fracture or flow in the impermeable overlay will, in such conditions, result in an *artesian spring*. Sometimes a small area of impermeable material may exist in a large aquifer. This may happen through geological faulting or, for example, through a lens of clay occurring in otherwise sandy glacial drift. A small local water table, called a *perched water table* may result and this can often be a long way above the true phreatic surface.

Some of the modes of occurrence of groundwater described above are illustrated in Fig. 5.1.

5.2 Factors of influence

The flow of groundwater takes place in porous media. The pores through which movement takes place may be very small indeed and generally are between the limits 2 mm–0·02 mm. The movement is slow by the standard of surface runoff and the flow is usually *laminar*. The Reynolds number in flow of this kind is very low.

The factors of importance in the flow are

(1) the liquid—its density and viscosity
(2) the media through which the liquid moves
(3) the boundary conditions

The liquid normally is water, usually fresh but occasionally saline. Its temperature may vary in range 0°–30°C.

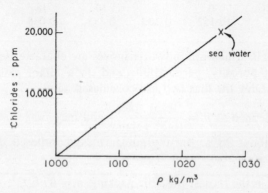

FIG. 5.2 *Density of saline water*

Density. The density of fresh water varies only very slightly with temperature and its effect may be ignored in groundwater flow. Of greater importance is salinity, the effect of which is shown in Fig. 5.2.

Density of Fresh Water v. *Temperature*

temperature °C	0	4	10	15	20
ρ g/l (grammes per litre)	999·868	1000	999·727	999·126	998·230

Viscosity is a measure of the shear strength of a liquid; the lower the viscosity, the more mobile or penetrating the liquid.

Absolute viscosity, denoted by μ, has units Ns/m^2; (the CGS unit, the *poise* = 10^{-1} Ns/m^2). Water at 20°C has a viscosity of 1 *centipoise* (0·01 poise = 10^{-3} Ns/m^2).

Kinematic viscosity, denoted by v, is the ratio of absolute viscosity to density, or $v = \dfrac{\mu}{\rho}$ and has units m²/s (the CGS unit, the *stoke* = 10^{-4} m²/s). Kinematic viscosity occurs in many applications, e.g. Reynolds number, $R = \dfrac{vD}{v}$.

$$v \text{ (water)} \approx 10^{-1} \text{ m}^2/\text{day} \approx 10^{-6} \text{ m}^2/\text{s} \approx 10^{-2} \text{ stoke.}$$

The kinematic viscosity of liquids is almost independent of pressure and is substantially a function of temperature.

Kinematic Viscosity of Water v. *Temperature*

temperature °C	0	5	10	15	20
v m²/day	0·152	0·133	0·113	0·098	0·087

The media in which groundwater moves are characterised by the properties of porosity, permeability and, to a minor extent compressibility. Only the first two are considered here.

Porosity is defined as $n = \dfrac{\text{total voids}}{\text{total volume}}$ and may range from a few percent to about 90%. In a granular mass composed of perfect uniform spheres:

in the loosest possible packing, $n = 47·6\%$

in the densest possible, $n = 26\%$

Natural soils are, of course, composed of irregular particles of many different sizes. The more regular the soil, the more porous it tends to be, since in irregular soils the smaller particles tend to fill the voids in the larger particle packing. It is, therefore, standard procedure in any groundwater survey to analyse the soils mechanically and plot the particle sizing in a standard way, using a logarithmic size scale. A typical analysis is shown in Fig. 5.3. Two soils are plotted: the more regular soil has the steeper slope and is likely to be more porous.

When water fills the pores of a soil there is a thin layer, only a few molecules thick, which coats the particles. This water is not free to move and adheres to the particles even when the voids have been drained, occupying part of the available space. This means that the *effective porosity*, n_e, may be less than the true porosity n. In coarse materials such as gravels there will be no difference between n_e and n but in fine sands the difference may be 5%, and even more in very fine materials. In most considerations of porosity in the flow of groundwater it is n_e, the effective porosity, which is of importance.

Permeability is a function of porosity, structure, and the geological history of the material. By structure is meant the grain size, distribution, orientation, arrangement and shape of the particles. For

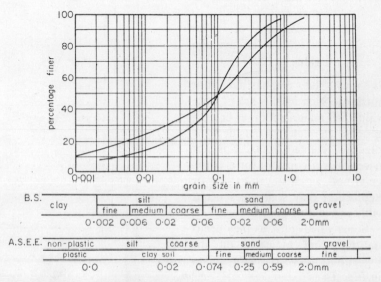

FIG. 5.3 *Plotting of particle size analysis, and nomenclature*

example in a sediment of predominantly flat grains, deposited in water, the grains will tend to lie with their longest axes horizontal. In such a soil the permeability may well be higher in a horizontal direction than in a vertical one. Such a soil is called *anisotropic*.

The permeability of a particular material is defined by its *permeability coefficient*, denoted by k; k depends on the factors listed above which may be described as the geometry of the pore system, and is expressed in metres per day, or feet per day.

Many attempts have been made to find a formula for k in terms of measurable values of the properties of the material. Generally this is very difficult and such formulae can only be used with confidence in strictly limited applications. For example, a formula used in connection with sand filters for water supply, and applying only to homogeneous rounded grain media, of not too fine a size, is

$$k = Cd_{10}^2$$

where k = permeability coefficient in m/day

d_{10} = the grain size in mm where 10% of material is finer and 90% is coarser

C = a constant of value 400–1200 (an average value is 1000)

It must be emphasised that formulae such as this are of little value in materials of a heterogeneous character or outside their precisely defined limits, k is not necessarily a constant for a particular soil, as chemical erosion or deposition may sometimes occur with percolating groundwater.

Some values of k as they occur naturally are indicated on the logarithmic scale of permeability in Fig. 5.4.

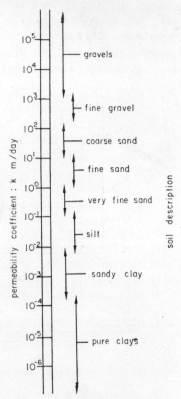

FIG. 5.4 *Range of permeability in natural soils*

5.3 Groundwater flow

5.3.1 Darcy's Law. Before any mathematical treatment of the flow of groundwater can be attempted, it is necessary to make certain simplifying assumptions

(1) the material is homogeneous and isotropic

(2) there is no capillary zone

(3) there is a steady state of flow

The fundamental law is Darcy's Law of 1856. This states that the rate of flow per unit area of an aquifer is proportional to the gradient of the potential head measured in the direction of flow. Or

$$v \propto i$$

By introducing a constant of proportionality, which is k, the permeability coefficient

$$v = ki$$

and for a particular aquifer or part of an aquifer of area A, (area at right angles to flow)

$$Q = vA = kAi$$

where $v =$ velocity of the water, (measured as the time taken to pass between 2 reference points) in m/day (or m/s or ft/s etc.) and called the *specific velocity*.

$i =$ the hydraulic gradient. This equals the potential gradient since velocities are so small there is virtually no velocity head. i is also written as $\dfrac{\mathrm{d}\phi}{\mathrm{d}s}$, s being distance along a flow line, and ϕ the potential head.

The specific velocity is not the true velocity, but is merely Q/A. The actual velocity of water in the pores is greater than the specific velocity since the path the water follows through porous media is always longer than a straight line between any two points.

If the average real, or pore, velocity is denoted by $\bar{v}$ then

$$\bar{v} = \frac{\text{discharge}}{\text{area of water passage}} = \frac{Q}{An_e} = \frac{Av}{An_e} = \frac{v}{n_e}$$

∴ pore velocity (average) = specific velocity ÷ effective porosity

Since the velocity distribution across a pore is probably parabolic, being highest in the centre and zero at the edges, the maximum pore velocity v max = approx. $2 \times \bar{v}$.

So in a typical case (say) $n_e = \frac{1}{3}$

then $$\bar{v} = 3v \quad \text{and} \quad \bar{v}_{\max} = 6v$$

While these are only typical figures it is important to remember the order of the velocities, since it is on $\bar{v}_{\max}$ that the Reynolds number and the continuance of laminar flow depends.

5.3.2 Flow in a confined aquifer. Consider now the case of unidirectional flow in a confined aquifer of permeability k, illustrated in Fig. 5.5. Groundwater is flowing from left to right, the energy required to move the water through the pores is continually using up the available head and so the line of potential head as indicated by piezometers introduced into the aquifer, is declining.

From Darcy's Law $v_x = -k \dfrac{\mathrm{d}\phi}{\mathrm{d}x}$

and if $q =$ flow in the aquifer per unit width then

$$q = -kH \frac{\mathrm{d}\phi}{\mathrm{d}x} \tag{5.1}$$

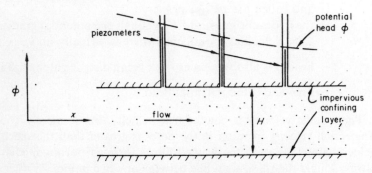

FIG. 5.5 *Flow in a confined aquifer*

Since it is assumed that the flow is a steady state

$$\frac{\mathrm{d}q}{\mathrm{d}x} = 0$$

and differentiating the equation for q, above

$$\frac{\mathrm{d}q}{\mathrm{d}x} = -kH \frac{\mathrm{d}^2\phi}{\mathrm{d}x^2}$$

and so

$$\frac{\mathrm{d}^2\phi}{\mathrm{d}x^2} = 0 \tag{5.2}$$

since both k and H have finite values.

These two equations (5.1) and (5.2) are the fundamental differential equations for this case of a confined aquifer. By introducing suitable boundary conditions, many problems in this case may be solved by the equations.

Note that the v_x in the Darcy equation is the specific velocity.

5.3.3 Flow in an aquifer with phreatic surface. Consider now the case of an aquifer with a phreatic surface resting on an impermeable

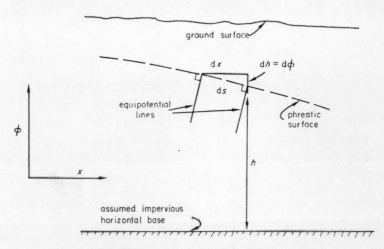

FIG. 5.6 *Flow in an aquifer with phreatic surface*

base. This case is illustrated in Fig. 5.6. Here the first equation from Darcy's Law, would be

$$v_s = -k\frac{\mathrm{d}\phi}{\mathrm{d}s}$$ where s = distance measured in the direction of flow.

If two assumptions (due to Dupuit) are made

1. that $\dfrac{\mathrm{d}\phi}{\mathrm{d}s}$ can be represented by $\dfrac{\mathrm{d}\phi}{\mathrm{d}x}$ (permissible if $\mathrm{d}\phi$ is small)
2. that all flow lines in the aquifer are horizontal and equipotential lines vertical (nearly true except near abstraction points) so that $\dfrac{\mathrm{d}\phi}{\mathrm{d}x} = \dfrac{\mathrm{d}h}{\mathrm{d}x}$ then the Darcy equation becomes

$$q = -kh\frac{\mathrm{d}h}{\mathrm{d}x} \qquad (5.3)$$

and

$$\frac{dq}{dx} = -\tfrac{1}{2}k \frac{d^2(h^2)}{dx^2}*$$

and since by continuity $\frac{dq}{dx} = 0$, then

$$\frac{d^2(h^2)}{dx^2} = 0 \qquad (5.4)$$

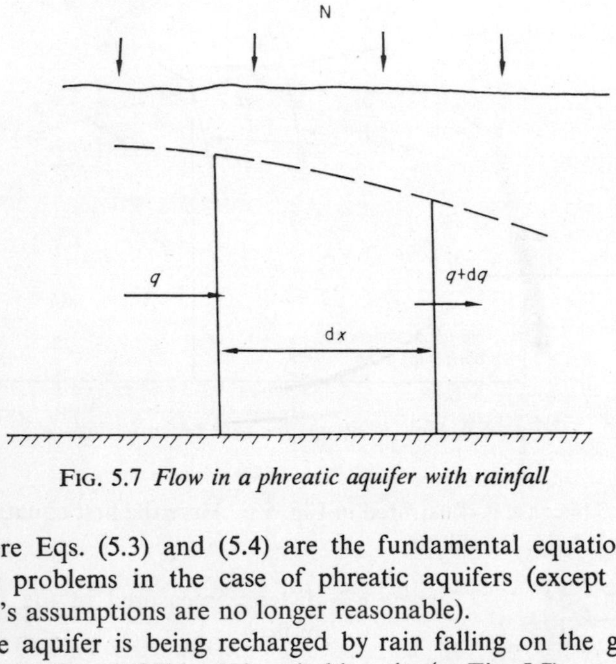

N

FIG. 5.7 *Flow in a phreatic aquifer with rainfall*

and here Eqs. (5.3) and (5.4) are the fundamental equations for solving problems in the case of phreatic aquifers (except where Dupuit's assumptions are no longer reasonable).

If the aquifer is being recharged by rain falling on the ground surface, let the rainfall be N in suitable units (as Fig. 5.7).

* The power rule of differentiation may be written $\frac{d(v^n)}{dx} = nv^{n-1}\frac{dv}{dx}$

$$\therefore \frac{d(h^2)}{dx} = 2h.\frac{dh}{dx}$$

$$\therefore \tfrac{1}{2}\frac{d(h^2)}{dx} = h.\frac{dh}{dx} \text{ and since } q = -kh\frac{dh}{dx}$$

$$\therefore q = -\tfrac{1}{2}k\frac{d(h^2)}{dx}$$

$$\therefore \frac{dq}{dx} = -\tfrac{1}{2}k\frac{d^2(h^2)}{dx^2}$$

In this case $dq = N \cdot dx$

$$\therefore \quad \frac{dq}{dx} = -\tfrac{1}{2}k \frac{d^2(h^2)}{dx^2} = N$$

$$\therefore \quad \frac{d^2(h^2)}{dx^2} = -\frac{2N}{k} \tag{5.5}$$

and now Eqs. (5.3) and (5.5) are relevant.

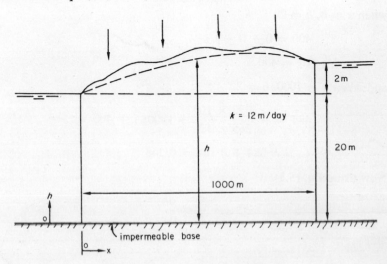

$k = 12 \, \text{m/day}$

h

$2 \, \text{m}$

$20 \, \text{m}$

$1000 \, \text{m}$

h

o

impermeable base

FIG. 5.8 *Flow between two canals*

Example 5.1. *Suppose there are two canals, at different levels, separated by a strip of land 1000 m wide, of permeability k = 12 m/day as shown in Fig. 5.8. If one canal is 2 m higher than the other and the depth of the aquifer is 20 m below the lower canal to an impermeable base, find the inflow into, or abstraction from, each canal per m length. Take annual rainfall as 1·20 m per annum and assume 60% infiltration.*

Assume a reference origin as indicated on the figure. Then the boundary conditions are simply

when $x = 0$, $h = 20$; and when $x = 1000$, $h = 22$.

$$N = 60\% \text{ of } 1 \cdot 2 = 0 \cdot 72 \text{ m/year}$$

$$= 0 \cdot 72/365 \text{ m/day}$$

From Eq. 5.5
$$\frac{d^2(h^2)}{dx^2} = -\frac{2N}{k}$$

Integrating twice $\dfrac{d(h^2)}{dx} = -\dfrac{2N}{k}x + C_1$

$$h^2 = -\frac{N}{k}x^2 + C_1 x + C_2 \qquad (5.6)$$

when $x = 0$, $h = 20$

$$\therefore \quad 400 = 0 + 0 + C_2$$

$$\therefore \quad C_2 = 400$$

and when $x = 1000$, $h = 22$

$$\therefore \quad 484 = -\frac{0 \cdot 72 \times 10^6}{365 \times 12} + 1000C_1 + 400$$

$$\therefore \quad C_1 = 0 \cdot 084 + 0 \cdot 164 = 0 \cdot 248$$

Now using Eq. (5.3).

$$q = -kh\frac{dh}{dx}$$

and Eq. (5.6)

$$h = \sqrt{\left(-\frac{N}{k}x^2 + 0 \cdot 248x + 400\right)}$$

Let the expression under the square root sign $= u$, then $h = u^{\frac{1}{2}}$

$$\therefore \quad \frac{dh}{dx} = \frac{1}{2u^{\frac{1}{2}}}\left(-\frac{N}{k} \cdot 2x + 0 \cdot 248\right)$$

at $x = 0$

$$q = -ku^{\frac{1}{2}} \cdot \frac{1}{2u^{\frac{1}{2}}}(0 \cdot 248) = -6(0 \cdot 248) = -1 \cdot 49 \text{ m}^3/\text{day/m}$$

at $x = 1000$

$$q = -\frac{k}{2}\left(-\frac{2000 \times 0 \cdot 72}{365 \times 12} + 0 \cdot 248\right)$$

$$= -6(-0 \cdot 328 + 0 \cdot 248) = 6(0 \cdot 08) = 0 \cdot 48 \text{ m}^3/\text{day/m}$$

So there is discharge into both canals from the aquifer of $1 \cdot 49 \text{ m}^3/\text{day}$ to the lower, and $0 \cdot 48 \text{ m}^3/\text{day}$ to the upper, for each metre length of aquifer.

The foregoing simple cases will serve to show the way in which the movement of groundwater can be analysed. As conditions become more complex, the solutions become more difficult, but standard solutions have been computed for very many groundwater situations* and most conditions met with in nature can be analysed, at least, approximately.

5.4 The abstraction of groundwater

The simplest and oldest way of collecting groundwater is by digging a hole in the ground which penetrates the water table. If the quantity which can be taken from the hole is not adequate, then the hole must be extended either horizontally or vertically. Which method is chosen will depend on the local geo-hydrology.

If the hole is extended horizontally it becomes an open collecting ditch. Alternatively, it could be underground as a collecting tunnel. These horizontal collectors must be used if the aquifer thickness is small and if the drawdown due to abstraction has to be limited, as for example when a layer of fresh water overlies a layer of salt water.

The vertical extension of the hole makes it a dug or drilled well, or a borehole. This method can be used when the aquifer is of sufficient thickness and in any case when the aquifer is more than about 6 m below ground level. Dug wells are usually 1 m or more in diameter and so the shaft acts as a reservoir for a short-time, high-rate abstraction. The large diameter is also useful when the entrance velocity of water into the shaft has to be kept low, e.g. in fine grained sands.

The majority of wells sunk nowadays for water supply are drilled wells and these are commonly from 30–500 m deep. They are constructed by using drill bits which break the material at the bottom of the hole into small pieces so that they may be removed with other tools. Two principal methods are employed: percussion drilling and rotary drilling. The percussion method involves alternately

* For a comprehensive treatment of the flow of groundwater and its recovery the reader is referred to two companion volumes in this series, *Theory of Groundwater Flow* by A. Verruijt, and *Groundwater Recovery* by L. Huisman.

raising and dropping the tools in the borehole; in the rotary method, a rotating bit, cuts or abrades the hole bottom. Drilled wells can penetrate any materials from soft clay to hard rock up to depths of a kilometre or more.

As the well is drilled it is "cased" with steel piping to prevent wall collapse. At the bottom of the casing a well screen is constructed. This is the point where groundwater enters and the screen is necessary to prevent the washing in of fine particles and consequent clogging of the well bottom and pump. The screen should cause as small a loss of head as possible, be structurally strong, corrosion resistant and reasonably cheap. These requirements are to some extent contradictory, since the smaller the screen openings and thus the more effective they are in keeping out fine particles, the greater the head loss which will be caused.

Modern well screens are usually slotted with fine slots in a plastic material, though steel, copper, bronze, wood, vitrified clay and glass are used. Gravel packs are customarily placed around the screen to act as preliminary filters, and in some cases gravel packs of reducing diameter particles, placed in concentric rings, may be used with a simply perforated bottom pipe on the casing.

The construction of drilled wells, well screens and gravel packs and the techniques of well development and maintenance are beyond the scope of this book.

Once the water has entered the well it has to be pumped to the surface. Well pumps may be classified as reciprocating, rotating vertical shaft, jet and air lift pumps. Rotating vertical shaft may be either surface driven or be submersible, and may be centrifugal or rotary positive displacement.

By far the most common application is now the electric submersible centrifugal borehole pump, with the electric drive motor directly coupled to the pump stages in one long pump body which is placed near the bottom of the well. Such pumps are manufactured in sizes down to 100 mm diameter to supply heads up to 100 m or more if necessary. Such a 100 mm (4 in.) diameter pump would supply about 4 m^3 per hour while a 250 mm (10 in.) diameter pump might supply 30 times as much. A sketch of a typical installation is shown in Fig. 5.9.

5.5 The yield of wells

Formulae for the draw down curves of a single well may be derived from the conditions of flow previously discussed in Section 5.3. Only the two simplest cases are shown here, (1) steady flow to a

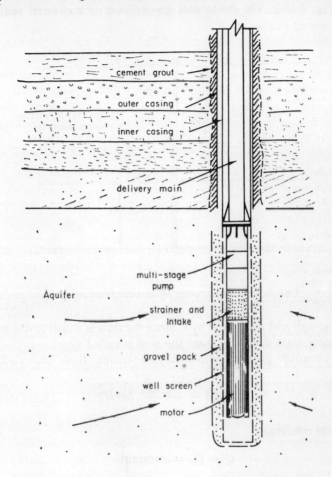

Fig. 5.9 *Sketch of bottom of a typical water supply borehole*

well tapping confined groundwater, and (2) steady flow to a water table well.

Many other factors should be considered, e.g. the influence of partial penetration of the well in the aquifer, unsteady flow etc., but for a full treatment of these the reader is referred to the companion volumes in this series previously mentioned.

5.5.1 Steady confined flow. Drawdown is denoted by ϕ and is measured from the undisturbed piezometric surface before pumping. (See Fig. 5.10.) The horizontal co-ordinate is measured radially

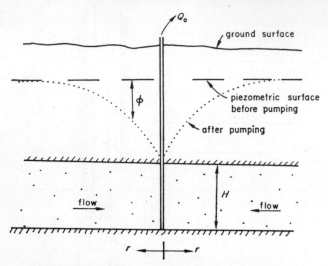

FIG. 5.10 *Well pumping from a confined aquifer*

from the well and is denoted by r, since the flow is radial to the well. The steady state discharge from the well is called Q_o.

The equations governing flow may be written: from Darcy's Law

$$Q = vA = -k\frac{\mathrm{d}\phi}{\mathrm{d}r} \cdot 2\pi r H$$

and from continuity

$$Q = Q_o = \text{constant}$$

combining these

$$\mathrm{d}\phi = -\frac{Q_o}{2\pi k H} \cdot \frac{\mathrm{d}r}{r}$$

Integrated between the limits $r = r_1$, $\phi = \phi_1$, and $r = r_2$, $\phi = \phi_2$

$$\phi_1 - \phi_2 = \frac{Q_o}{2\pi k H} \cdot \ln\frac{r_2}{r_1} \tag{5.7}$$

Indefinitely integrated, it yields

$$\phi = -\frac{Q_o}{2\pi kH} \cdot \ln r + C$$

and if $\phi = 0$ when $r = R$ then

$$\phi = \frac{Q_o}{2\pi kH} \ln \frac{R}{r} \qquad (5.8)$$

Either of these two equations (5.7) and (5.8) enables the drawn-down curve to be established if the boundary conditions are known.

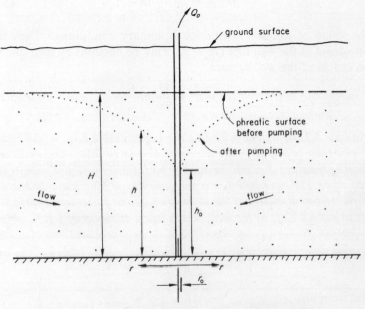

FIG. 5.11 *Well pumping from an unconfined aquifer*

5.5.2 Steady unconfined flow. When the drawdown is slight compared with the thickness of the aquifer, the factor kH remains nearly constant and the formula for steady confined flow may be used. As drawdown increases, the falling-water level reduces the area of transmitting aquifer and the equations now become, in the notation of Fig. 5.11, Darcy

$$Q = 2\pi r \cdot h \cdot k \frac{dh}{dr}$$

Continuity

$$Q = Q_o = \text{constant}$$

Combined

$$h \cdot dh = \frac{Q_o}{2\pi k} \cdot \frac{dr}{r}$$

Integrated

$$h^2 = \frac{Q_o}{\pi k} \ln r + C \qquad \text{and if } h = H \text{ at } r = R$$

$$H^2 - h^2 = \frac{Q_o}{\pi k} \ln \frac{R}{r} \qquad (5.9)$$

The value of R must satisfy the boundary conditions. Then the drawdown at the well face $(H - h_0)$ may be deduced by introducing the radius of the well r_0 since

$$H^2 - h^2 = \frac{Q_o}{\pi k} \ln \frac{R}{r_0} \qquad (5.10)$$

Example 5.2. *A well is drilled to the impermeable base in the centre of a circular island of* 1 *mile diameter in a large lake. The well completely penetrates a sandstone aquifer* 50 *ft thick overlain by impermeable clay. The sandstone has a permeability of* 50 *ft/day. What will be the steady discharge if the drawdown of the piezometric surface is not to exceed* 10 *ft at the well, which has a diameter of* 1 *ft?*

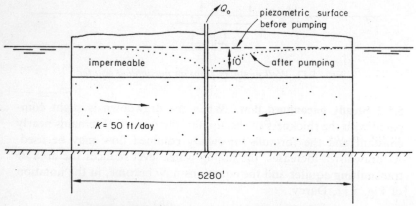

FIG. 5.12 *Pumping from a central well in a circular island and confined aquifer*

For a well in the centre of the circular island, the boundary condition is $\phi = 0$ when $r = 2640$ ft.

Then from Eq. (5.8)

$$10 = \frac{Q_o}{2\pi \times 50 \times 50} \ln \frac{2640}{0 \cdot 5}$$

$$\therefore \quad Q_o = \frac{50000\pi}{\ln 5280} = \frac{157080}{8 \cdot 572} = 18340 \text{ ft}^3/\text{day}$$

$$= 0 \cdot 212 \text{ ft}^3/\text{s}$$

5.5.3 Steady unconfined flow with rainfall.

When rainfall is present the equations become

Darcy $$Q = 2\pi r h \cdot k \frac{dh}{dr}$$

Continuity $$dQ = -2\pi r \cdot dr \cdot P$$

integrating $\quad Q = -\pi r^2 P + C_1$ and C_1 may be determined from the condition that when $r = r_o \approx 0 \qquad Q = Q_o$

$$\therefore \quad Q = -\pi r^2 P + Q_o$$

Substituting this value in the Darcy equation

$$h \cdot dh = \frac{Q_o}{2\pi k} \cdot \frac{dr}{r} - \frac{P}{2k} \cdot r \, dr$$

and integrating $$h^2 = \frac{Q_o}{\pi k} \ln r - \frac{P}{2k} r^2 + C_2 \qquad (5.11)$$

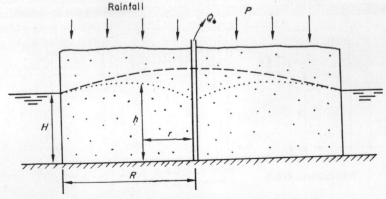

FIG. 5.13 *Central well in a circular island and unconfined aquifer with rainfall*

C_2 being an integration constant that must satisfy the particular boundary conditions. In the case of a central well in a circular island of radius R when $r = R$, $h = H$ and so

$$C_2 = H^2 - \frac{Q_o}{\pi k} \ln R + \frac{2k}{P} R^2$$

substituting this value in Eq. (5.11);

$$H^2 - h^2 = \frac{Q_o}{\pi k} \ln \frac{R}{r} - \frac{P}{2k}(R^2 - r^2) \qquad (5.12)$$

If $Q_o = 0$, i.e. there is no pumping, then the shape of the phreatic surface is given by

$$H^2 - h^2 = -\frac{P}{2k}(R^2 - r^2) \qquad (5.13)$$

Example 5.3. A circular island 500 m radius has an effective rainfall $P = 4$ mm/day. A central well is pumped to deliver a constant $Q_o = 25$ m³/h from an aquifer with dimensions and properties as shown in Fig. 5.14. What is the drawdown at the well and at the water divide?

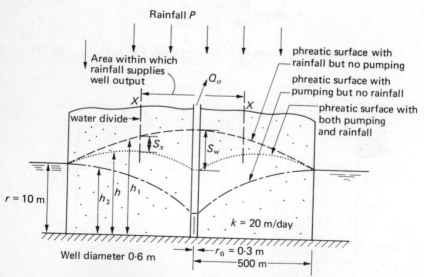

FIG. 5.14 *Circular island with central well, rainfall and an unconfined aquifer. Solution by super position*

(1) Assume no pumping. Then from Eq. (5.13)

$$H^2 - h_1^2 = -\frac{P}{2k}(R^2 - r^2)$$

$$100 - h_1^2 = -\frac{0 \cdot 004}{40}(250000 - r^2)$$

$$100 - h_1^2 = -25 + \left(\frac{r}{100}\right)^2 \tag{5.14}$$

(2) Assume no rainfall. Then from Eq. (5.9)

$$H^2 - h_2^2 = \frac{Q_o}{\pi k} \ln \frac{R}{r}$$

$$100 - h_2^2 = 21 \cdot 9 \log \frac{500}{r} \tag{5.15}$$

Now applying the method of super position and combining Eqs. (5.14) and (5.15)

$$100 - h^2 = -25 + \left(\frac{r}{100}\right)^2 + 21 \cdot 9 \log \frac{500}{r} \tag{5.16}$$

when $r = 0 \cdot 3$ m (at well face) $h = 7 \cdot 39$

and from Eq. (5.14) $h_1 = 11 \cdot 19$ and calling drawdown at the well S_w ∴ $S_w = 3 \cdot 80$ m below the phreatic surface before pumping.

If a water-divide exists, then all the output of the well is being contributed by rainfall, since if the sea around the island were contributing, the hydraulic gradient would be sloping downwards and inwards at every point. So the area contributing is obtained from

$$Q_o = \pi r_x^2 P \quad \text{where } r_x \text{ is radius of the divide}$$

$$25 = \pi r_x^2 \times \frac{0 \cdot 004}{24}$$

∴ $r_x = 218$ m (well within the 500m radius of the land)

using this value in the combined Eq. (5.16)

$$h = 10 \cdot 61$$

and from Eq. (5.14) $h_1 = 10.97$, and calling drawdown at the divide S_x

$$S_x = 0.36 \text{ m}$$

Suppose the simple formula of Eq. (5.8) had been used, thus assuming constant aquifer thickness. In this case

$$S_w = \frac{Q_o}{2\pi k H} \ln \frac{R}{r} = \frac{600}{2\pi \times 20 \times 10} \times 2.3 \log \frac{500}{0.3}$$

$$= 0.477 \times 2.3 \times 3.223$$

$$= 3.54 \text{ m (cf 3.80 m above)}$$

and

$$S_x = 0.477 \times 2.3 \times \log \frac{500}{218}$$

$$= 1.01 \times 0.36$$

$$= 0.364 \text{ (cf 0.36 m above)}$$

It will be realised that the simple formula for the confined case is adequate in this case except in the immediate neighbourhood of the well. It would, of course, still be necessary to compute the 'no-pumping' phreatic surface.

6 Surface Runoff

6.1 The engineering problem

Rainfall, if it is not intercepted by vegetation or by artificial surfaces such as roofs or pavements, falls on the earth where it may evaporate, infiltrate or lie in depression storage. When the losses arising in these ways are all provided for, there may remain a surplus which, obeying the gravitation laws, flows over the surface to the nearest stream channel. The streams coalesce into rivers and the rivers find their way down to the sea. When the rain is particularly intense or prolonged, or both, the surplus *runoff* becomes large and the stream and river channels cannot accept all the water suddenly arriving. They become filled and overflow and in so doing they do great harm to the activities of men. The most serious effect of flooding may be the washing away of the fertile top soil in which crops are grown, and of which there is already a scarcity on the Earth. In urban areas there is great damage to property, pollution of water supplies, danger to life and often total disruption of communications. In agrarian societies floods are feared like pestilence because they may destroy crops, cattle and habitations, and bring famine in their wake.

The hydraulic engineer in dealing with runoff, has to try to provide answers to many questions, of which some of the more obvious are

  (i) how often will floods occur?
 (ii) how large will they be and to what level will they rise?
(iii) how often will there be droughts?
(iv) how long may these droughts continue?

Questions of this kind are all variations of one, which is concerned with the magnitude and duration of runoff from a particular catchment with respect to time. They can be resolved only by the determination of the frequency and duration of particular discharges from observations over a long period of time, though if such observations are not available, estimations may still be made at various probabilities.

A second group of questions arises in using the curves of runoff frequency and duration once found, e.g.

(i)   how can the volume of discharge be reduced?
(ii) how can the cost of flood prevention be compared with the damage which will arise if no measures are taken?
(iii) how valuable is stored flood water in times of drought?

These questions are not directly related and each involves a quite different and distinct approach, though the same techniques may be used in answering more than one. In this and the following sections ways in which some of these questions may be answered will be sought.

6.2 Flow rating curves: their determination adjustment and extension

6.2.1 Definition. A rating curve is a graph drawn connecting the water level elevation, or *stage* of a river channel at a certain cross-section with the corresponding discharge at that section. A typical

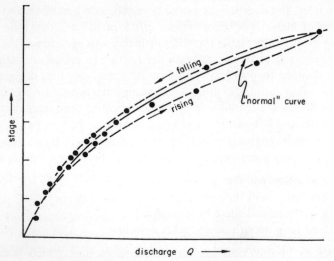

Fig. 6.1 *Flow rating curve*

rating curve is shown in Fig. 6.1. It may be seen that the curve is drawn through a cluster of points, each of which represents the results of a river discharge measurement. Such measurements may be made in a number of ways of which the most important are

(1) by velocity–area methods
(2) by flow-measuring structures
(3) by dilution gauging

6.2.2 Velocity–area methods. These are conventional for medium to large rivers and involve the use of a *current meter* which is a fluid velocity measuring instrument. A small propeller rotates about a horizontal shaft which is kept parallel to the streamlines by tail fins. The instrument is ballasted to keep it as nearly as possible directly below the observer. Another version of the instrument has a circlet of small conical cups disposed horizontally about the suspension axis.

Each revolution of the propeller is recorded electrically through a cable to the observer and the number of revolutions is counted by the observer or automatically over a short period (say 1 or 2 minutes). These observations are converted into water velocities from a calibration curve for the instrument. By moving the meter vertically and horizontally to a series of positions whose co-ordinates in the cross-section are known a complete velocity map of the section may be drawn and the discharge through the cross-section calculated. Fig. 6.2 shows a modern current meter assembled for use on its supporting cable which is also used as a depth measure.

Observations may be made by lowering the meter from a bridge, though if the bridge is not a single-span one, there will be divergence and convergence of the streamlines caused by the piers which can cause considerable errors. In many instances, however, the gauging site, which should be in as straight and uniform a reach of the river as is possible, will have no bridge, and if it is deep and in flood a cable to hold some stable boat must be provided, together with a lighter measuring cable to determine horizontal position in the cross-section.

Since the drag on a boat with at least two occupants and suspended current meter is considerable, a securely fastened steel cable should be used. The propinquity of suitable large trees at a particular cross-section may often necessitate its choice for this reason.

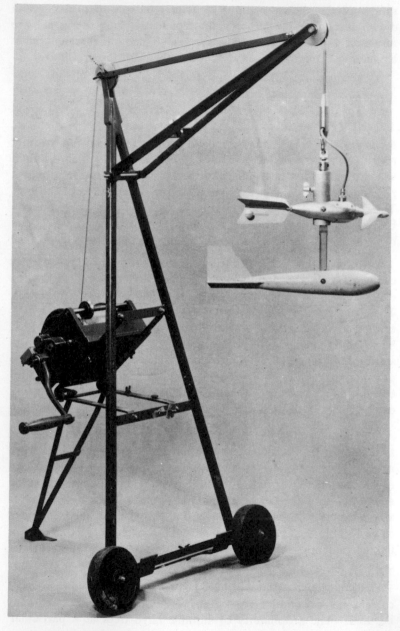

FIG. 6.2 *A modern helix current meter by Hilger and Watts Ltd*

Alternatively, cableways are sometimes used to suspend the meter, either from a manned cable-car or directly from the cable, the instrument in this latter case being positioned by auxiliary cables from the river bank.

Depths should always be measured at the time of velocity observation since a profile can change appreciably during flood discharges. Observers should also remember such elementary rules as to observe the stage before and after the discharge measurement, and to observe the water slope by accurate levelling to pegs at the river level as far up and downstream of the gauging site as is practicable, up to say 500 m in each direction.

As water velocities increase in high floods the ballasted current-meter will be increasingly swept downstream on an inclined cable. The position of a meter in these circumstances may be found reasonably accurately if the cable angle is measured. Ballast may be increased but only within limits. Rods may be used to suspend the meter but a rigid structure in the boat will then be required to handle the rods, calling for a stable platform of the catamaran type. Rod vibration and bending are common in deep rivers unless diameters exceed 50 mm by which time the whole apparatus is getting very heavy and unmanageable.

It will be appreciated that since each river is unique each will require a careful assessment of its width, depth, likely flood velocities, cable support facilities, availability of bridges, boats, etc. before a discharge measurement programme is started.

From many observations on many rivers it has been established that the variation of velocity integrated over the full depth of a river can be closely approximated by the mean of two observations made at 0.2 and 0.8 of the depth. If time and circumstances preclude even two observations at each horizontal intercept then one reading at 0·6 depth will approximate the average over the whole depth.

The discharge at the cross-section is best obtained by plotting each velocity observation on a cross-section of the gauging site with an exaggerated vertical scale. *Isovels* or contours of equal velocity are then drawn and the included areas measured by planimeter. A typical cross-section, so treated, is shown in Fig. 6.3. Alternatively the river may be subdivided vertically into sections and the mean velocity of each section applied to its area.

A check should always be made using the slope–area method of Section 6.2.6(iii) and a value obtained for Manning's n. In this way a knowledge of the n values of the river at various stages will be built up which may prove most valuable in extending the discharge rating curve subsequently.

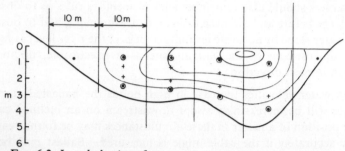

FIG. 6.3 *Isovel plotting of stream cross-section* (*points ringed are those for* 0·2 *and* 0·8 *depth observations*)

To ensure uniformity in the techniques of current-meter gauging the International Organisation for Standardisation (I.S.O.) has published various recommendations and, in the United Kingdom, BS 3680 "Measurement of liquid flow in open channels" refers.

6.2.3 Flow-measuring structures. These are designed so that stream discharge is made to behave according to certain well-known hydraulic laws. For example, the discharge per unit length over a weir is a function of the head over the weir. Many specialised weirs, such as V-notch, compound and Crump weirs, have been designed to provide accurate discharge data by observations of water surface level upstream of the weir. Similarly with flumes, where a stream is channelled through a particular geometrically shaped regular channel section for some distance before entering a length of different cross-section, made so usually by side contractions or steps in the bed. Trapezoidal shapes are often used and, recently, narrow vertical sections have been employed for catchment discharge measurement in Wales. Fig. 6.4 shows a flume of the latter type on the Plynlimon experimental catchments of the Institute of Hydrology. Generally, however, flow-measuring structures are confined to streams and fairly small rivers, since for large flows and wide rivers they become extremely expensive to build.

6.2.4 Dilution gauging. Dilution gauging is particularly suited to small turbulent streams where depths and flows are inappropriate for current-metering and flow-structures would be unnecessarily expensive and/or permanent.

The method involves the injection of a chemical into the stream and the sampling of the water some distance downstream after complete mixing of the chemical in the water has occurred. The chemical can either be added by *constant-rate injection* until the

FIG. 6.4 *Narrow flow-measurement flume Plynlimon catchments*

sampling downstream reveals a constant concentration level, or administered in a single dose as quickly as possible known as *gulp injection*. In this case samples over a period of time disclose the concentration/time correlation. In both cases the concentration of chemical in the samples is used to compute the dilution, and hence the discharge of the stream may be obtained. Fig. 6.5 shows constant-rate injection of sodium dichromate from a Mariotte bottle (a constant-head device) in a mountain stream.

Analysis of the samples is by an automated colorimetric procedure which estimates the concentration of very small amounts of the chromium compound by comparison with a sample of the injection solution. The equipment is expensive and specialised.

The literature on dilution gauging is not large but references [35] and [36] will provide excellent guidance.

The methods of Sections 6.2.2, 6.2.3, and 6.2.4 used singly or in conjunction, will establish the correlation for any stream or river discharge with stage.

A rating curve when established in this way, enables a single observation of stage, made each day at a set time, by an unskilled observer, to be converted into a discharge rate and hence into a finite quantity of water, flowing at the observation point. The difficulty about rating curves is to obtain enough points at times of high discharge to enable an accurate correlation to be obtained.

6.2.5 Rating-curve adjustment. So far rating curves have been discussed in terms which imply that they are simply median lines drawn through a scatter of observation points. This is not the case. If each discharge point has been noted as being measured on a 'falling' or 'rising' stage the curve would strictly speaking become a loop as shown by the dashed line in Fig. 6.1.

This variation, or looping effect is due to several causes.

The first of these is *channel storage*. As the surface elevation of a river rises water is temporarily stored in the river channel.

Example 6.1. *Suppose that the gauge shows a rise at the rate of* 0·2 *m/h during a discharge measurement of* 100 *m³/s and the channel is such that this rate of rise may be assumed to apply to a* 1000 *m reach of river between the measurement site and the reach control.**

Let the average width of channel in the reach be 100 m. Then the rate of change of storage in the reach, dS is given by

$$dS = 1000 \times 100 \times 0.2$$
$$= 20{,}000 \text{ m}^3/\text{h}$$
$$= 5.6 \text{ m}^3/\text{s}.$$

The discharge measurement should be plotted on the rating curve as 94·4 m³/s (not as 100), since this is the discharge past the *control* corresponding to the mean gauge height.

The second reason for the looping of rating curves is the variation in surface slope that occurs as a flood wave moves along the channel.

* The *control* of a river reach is the section at which the profile of the river changes. For a full treatment of river profiles and controls, and discharge measurements, see *Flow in Channels* by R. H. J. Sellin, a companion volume in this series.

FIG. 6.5 *Dilution gauging: dispensing sodium dichromate solution from a Mariotte bottle*

Fig. 6.6 represents a longitudinal section of a flood wave passing along a river channel. As point a passes the gauging site, the gauge reads h, the river cross-section is A and the slope of the water surface s_1. When the flood wave has moved on so that point b is at the gauge site, the gauge reading h and the cross-section A are the same but the slope is now s_2. From the Manning formula

$$Q = Av = \frac{AR^{\frac{2}{3}}S^{\frac{1}{2}}}{n}$$

and so two different discharges will occur in the two cases since s changes while A, R and n remain constant.

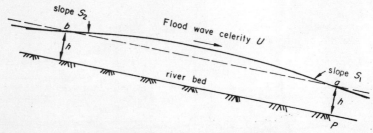

FIG. 6.6 *Flood wave slope variation*

Since the rising stage is associated with the greater slope discharge, measurements taken on a rising stage will plot to the right on the rating curve of Fig. 6.1 and those on a falling stage, to the left. Depending on the 'peakiness' of the flood wave, it often happens that maximum discharge occurs before the maximum stage is reached, since the influence of the steeper slope on velocity may outweigh the slight increase in cross-sectional area.

It is generally necessary to correct the discharge measurements taken on either side of flood waves to the theoretical steady state condition, because the great majority of gauge readings are taken by unskilled observers once a day without any indication of whether the stage is rising or falling. By using a corrected, or steady state, curve, the rising and falling stage observations will balance in the long run, and no value judgment or second daily visit to the gauge will be required from the observer. The correction may be made as follows.

From the Manning equation, the steady state discharge Q in a channel of given roughness and cross-section, is given as

$$Q \propto \sqrt{S} \qquad (6.1)$$

where S = steady state slope.

When the slope is not equal to S, as is the case in conditions of rising or falling stage, the actual discharge Q_a is given by

$$Q_a \propto \sqrt{(S \pm \Delta S)} \qquad (6.2)$$

Referring now to Fig. 6.7, ΔS may be expressed in terms of the rate of change in stage and flood wave celerity, U.

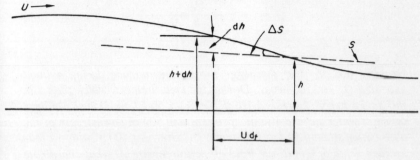

FIG. 6.7 *Change in slope of a flood wave with time*

The figure represents an advancing flood wave and rising stage. The gauge reading at the commencement of the discharge measurement is h and at its conclusion, dt, later, $h + dh$. In this time the wave has advanced $U dt$, and

$$\Delta S = \frac{dh}{U dt} = \frac{dh/dt}{U} \qquad (6.3)$$

$\frac{dh}{dt}$ being positive for a rising stage, negative for falling stage. Combining Eqs (6.1), (6.2) and (6.3)

$$\frac{Q_a}{Q} = \sqrt{\left(\frac{S + (dh/dt)/U}{S} \right)} \qquad (6.4)$$

or

$$\frac{Q_a}{Q} = \sqrt{\left(1 + \frac{dh/dt}{US} \right)} \qquad (6.5)$$

If the discharge measurement is taken at a site with two gauges, one upstream and one downstream, in the same reach, then all the terms in Eq. (6.5) are measured except Q and U, Q being the steady state discharge required and U, the flood wave celerity. There are now several ways to proceed. The first is to take an empirical value for U. Corbett [37] suggests that in a fairly uniform channel, in flood conditions, the celerity U is approximately equal to 1·3 times the mean water velocity, or

$$U = 1 \cdot 3 \frac{Q_a}{A}$$

where A = cross-sectional area.

From which

$$Q = \frac{Q_a}{\sqrt{\left(1 + \dfrac{A \cdot dh/dt}{1 \cdot 3 Q_a S}\right)}} \qquad (6.6)$$

Example 6.2. A river discharge measurement made during a flood indicated $Q_a = 3160 \ m^3/s$. During the measurement, which took 2 h, the gauge height increased from 50·40 to 50·52 m. Level readings taken at water surface 400 m upstream and 300 m downstream of the observation site differed by 100 mm. The river was 500 m wide with an average depth of 4 m at the time of measurement. At what coordinate should the measurement be plotted on the rating curve?

The cross-sectional area of the river, A = 500 m × 4 m

$$= 2000 \ m^2$$

$$\therefore \quad \text{average water velocity} = \frac{Q_a}{A} = \frac{3160}{2000} = 1 \cdot 58 \ m/s$$

Assume the flood wave celerity $U = 1 \cdot 3 \times 1 \cdot 58 = 2 \cdot 054$ m/s $\dfrac{dh}{dt} = \dfrac{0 \cdot 12 \ m}{7200 \ s} = 1 \cdot 67 \times 10^{-5}$ and $S = \dfrac{0 \cdot 1}{700} = 1 \cdot 43 \times 10^{-4}$. Then for a rising river, from Eq. (6.6)

$$Q_{corr} = \frac{3160}{\sqrt{\left(1 + \dfrac{1 \cdot 67 \times 10^{-5}}{2 \cdot 054 \times 1 \cdot 43 \times 10^{-4}}\right)}} = \frac{3160}{\sqrt{(1 \cdot 057)}} = 3080 \ m^3/s$$

and taking the mean gauge height, the corrected coordinates are 50·46 m and 3080 m³/s.

An alternative procedure, due to Boyer [38] is available, where values for neither U, nor S, need be obtained. If a sufficient number of observations are available, including measurements taken during rising and falling stages and in steady states, then a rating curve may be drawn as a median line through the uncorrected values. The steady state discharge Q can now be estimated from the median curve. Since Q_a and dh/dt are measured quantities and therefore known, Eq. (6.5) yields the term $1/US$ for each measurement of discharge.

The term $1/US$ is now plotted against stage and an 'average' curve fitted to the plotted points, as shown in Fig. 6.8. From the $1/US$

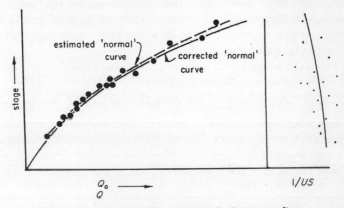

FIG. 6.8 *Method of correcting discharge readings without computing U or S*

v. stage relationship, new values of $1/US$ are obtained and inserted in Eq. (6.5) to yield the steady state Q. The new values of Q are then plotted v. stage as the corrected, steady-state curve.

Another method, which uses the observed slope but avoids evaluating U is due to Mitchell [39].

6.2.6 Extension of rating curves. As previously mentioned, the most difficult measurements of discharge to take are those at high flood, because of both the physical difficulties of high water velocity and floating debris, and the rare occurrence of the condition. It frequently happens, for example, that the condition for which river structures such as dams and bridges must be designed are defined as being "such that occur no more often than once in 100 years". This means that the structural designers want to know the probable discharge that will occur, on average, once every 100 years. This is sometimes referred to as "the 100 year flood".

If discharge measurements have been taken throughout the previous 100 years this design flood may not be too difficult to find. However, in the vast majority of cases, only stage readings will be available and then only for limited periods. If the hydraulic engineer can have 30 years of continuous daily stage readings and discharge measurements to include even low flood conditions, then he is fortunate. Almost always he will have to extend the rating curve well beyond the last measured point, to estimate the discharge at particular stage levels recorded. Occasionally the stage levels reached in high flood are recorded only by lines of debris on the banks, or grasses caught in the branches of riparian trees and scrub. Such physical evidence is valuable.

There are a number of ways of attempting rating curve extension:
(i) *By fitting an equation to the curve.*

Usually an equation of the form $Q = k(h - a)^x$ is used, where

h = stage

k and x are constants derived from the observed portion of the curve, and

a = height in m (or ft) between zero on the gauge and the elevation of zero flow.

Such a curve plots as a straight line on logarithmic paper and so may be easily extended. At best it is a questionable procedure since there is little theoretical justification for exponential laws operating and at high flows there may be quite abrupt changes in cross-section with rising stage.

(ii) *Steven's method* [40]. This method is based on Chezy's formula

$$Q = AC\sqrt{(RS)}$$

where A = cross-sectional area

C = Chezy roughness coefficient

R = hydraulic radius

S = slope of the energy line

If $C\sqrt{S}$ is assumed constant and D, the mean depth substituted for R then

$$Q = kA\sqrt{D} \qquad (6.7)$$

Known values of $A\sqrt{D}$ and Q are plotted and often come close to a straight line which may be extended. Field values of $A\sqrt{D}$ above the measured rating may then be used from the extended line to plot Q v. stage points on the rating curve.

The objection to this method is simply that $C\sqrt{S}$ is not a constant. However, as it takes account of the varying stream geometry it is a more rational procedure than (i).

(iii) *Slope area method.* This method depends on hydraulic principles and presupposes that it is practically possible to drive in pegs or make other temporary elevation marks at the time of the peak flow upstream and downstream of the discharge measuring site. These marks may subsequently be used to establish the water slope. Cross-sectional measurements will yield the area and hydraulic radius of the section. Then from Manning

$$Q = \frac{A R^{\frac{2}{3}} S^{\frac{1}{2}}}{n} \tag{6.8}$$

This method is sometimes criticised because of its dependence on the value of n. Since n for natural streams is about 0.035 an error in n of 0.001 gives an error in discharge of 3%. This objection may be partially met by plotting n v. stage for all measured discharges, so that the choice of n for high stages is not arbitrary, but is taken from such a plot. If the high flood slope can be measured then this method is probably the best one.

It should be emphasised that all methods of rating curve extension are suspect to some degree and should only be resorted to if no flood measurements can be obtained. The latter two methods, above, are both subject to errors arising from alteration of cross-section due to flood scour and subsequent low water deposition, so cross-sections and mean depth measurements should be taken as near to the time of discharge measurement as possible.

6.3 Duration of runoff

While floods and droughts are important from many points of view, they tend, as extremes, to be of comparatively short duration. For many water resource investigations it is equally important to know the total volumes of water that have to be dealt with over long periods of time, e.g. in hydroelectric power generation the plant capacity must be chosen for some discharge well below the peak flood since otherwise much capacity would be almost permanently idle. For such purposes the most convenient means of presenting data are the *mass curve* and the *flow duration curve*.

With an adjusted and well measured rating curve, the daily gauge readings may be converted directly to runoff volumes. A typical set of such daily runoff figures is presented graphically in Fig. 6.9. Such

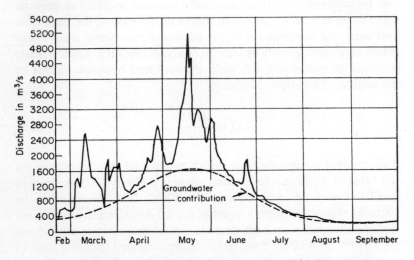

FIG. 6.9 *Hydrograph of R. Euphrates at Hit. Feb.–Sept. 1957 (after Directorate of Irrigation. Iraq)*

a presentation is called a *hydrograph*. If these volumes are plotted against time by adding each day's volume to the previous total, a cumulative mass curve of runoff is obtained. Such a curve is shown in Fig. 6.10.

Mass curves are extremely useful in reservoir design studies since they provide a ready means of determining storage capacity necessary for particular average rates of runoff and drawoff. Suppose for example that the mass curve *OA* of Fig. 6.11 represents the runoff from a catchment which is to be used for base load hydroelectric development. If the required constant drawoff is plotted on the same diagram, as line *OB*, then the required storage capacity to

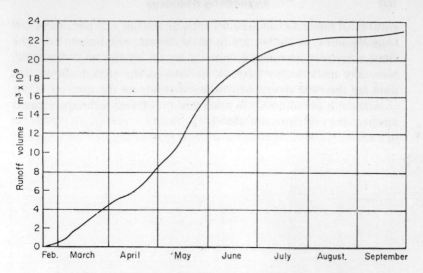

FIG. 6.10 *Cumulative mass curve of runoff for R. Euphrates at Hit. Feb.–Sept, 1957*

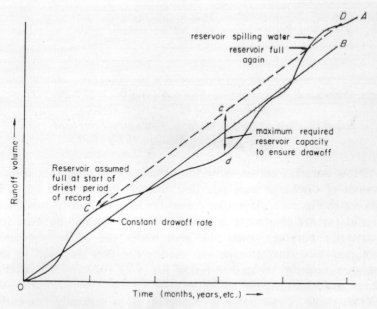

FIG. 6.11 *The use of mass curves in reservoir design*

ensure this rate may be found by drawing the line CD parallel to OB from a point C at the beginning of the dryest period recorded. The storage capacity necessary is denoted by the maximum ordinate cd. Normally much longer periods, as long as the record allows, are used for reservoir design and in many instances the drawoff is not constant nor continuous. In such cases, different techniques based on the same principles are used [41].

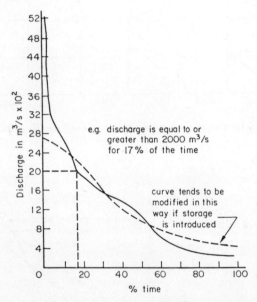

FIG. 6.12 *Flow duration curve for R. Euphrates, Feb.–Sept. 1957, at Hit (derived from the data of Fig. 6.9)*

Flow duration curves show the percentage of time that certain values of discharge were equalled or exceeded. Such a curve is shown in Fig. 6.12. Duration curves for long periods of runoff are useful for deciding what proportion of flow should be used for particular purposes, since the area under the curve represents volume. Note that storage will modify the flow duration in the manner shown by the dashed line of Fig. 6.12, reducing the extreme flows and increasing the minimum.

The shape of the curve is important as a generally flat curve indicates a river with few floods which is extensively supplied from groundwater, while a steep curve indicates a river with frequent floods and dry periods, having little groundwater flow and being supplied mainly from runoff.

The durations of particular floods or droughts, rather than the cumulative periods of particular discharge values, are also of interest and are usually obtainable directly from the hydrograph.

6.4 Catchment characteristics and their effects on runoff

It is now appropriate to consider how various properties of the *catchment area* affect the rate and quantity of discharge from it. By 'catchment area' is meant the whole of the land and water surface area contributing to the discharge at a particular stream or river cross-section, from which it is clear that every point on a stream channel has a unique catchment of its own, the size of catchment increasing as the control point moves downstream, reaching its maximum size when the control is at the sea coast.

There are many catchment properties which influence runoff and each may be present to a large or small degree. The intention, in analysing them separately, is to try to determine the effect of each characteristic on precipitation and its subsequent drainage from the catchment through the river channels.

(a) *Catchment area.* The area as defined at the beginning of this section is usually, but not necessarily, bounded by the topographic *water-divide.* Fig. 6.13 shows a hypothetical cross-section through the topographic water-divide of a catchment. Because of the underlying geology it is perfectly possible for areas beyond the divide to contribute to the catchment. The true boundary is indeterminate, however, because although some of the groundwater on the left of the divide in the figure may arrive in catchment B, the surface runoff will stay in catchment A. Here the infiltration capacity of the soil and the intensity of the rainfall will influence the portion of the rainfall that each catchment will collect.

If runoff is expressed, not as a total quantity for a catchment, but as a quantity per unit area, (usually m^3/s per square kilometre or ft^3/s per square mile) it is observed, other things being equal, that peak runoff decreases as the catchment area increases. This is due to the time taken by the water to flow through the stream

channels to the control section (the *time of concentration*) and also
to the lower average intensity of rainfall as storm size increases (see
Sect. 2.8.4.). Similarly, minimum runoff per unit area is increased
due to greater areal extent of the groundwater aquifers and minor
local rainfall.

(b) *Slope of catchment.* The more steeply the ground surface is
sloping the more rapidly will surface runoff travel, so that concentration times will be shorter and flood peaks higher. Infiltration

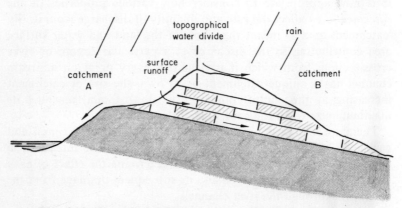

FIG. 6.13 *Watershed defined by geology as well as topography*

capacities tend to be lower as slopes get steeper, since vegetation is
less dense and soil more easily eroded, thus accentuating runoff.

Slope may be enumerated by covering a catchment contour map
with a rectilinear grid and evaluating the slope, perpendicular to
the contour lines at each grid point as shown in Fig. 6.14(a). A frequency distribution of these numbers may then be plotted as in
Fig. 6.14(b). Different catchments may be compared on the same
plot, the relatively steep frequency curves indicating catchments of
fast runoff and flat curves the converse.

(c) *Catchment orientation.* Orientation is important with respect to
the meteorology of the area in which the catchment lies. If the
prevailing winds and lines of storm movement have a particular
seasonal pattern, as they usually have, the runoff hydrograph will
depend to some degree on the catchment's orientation within the

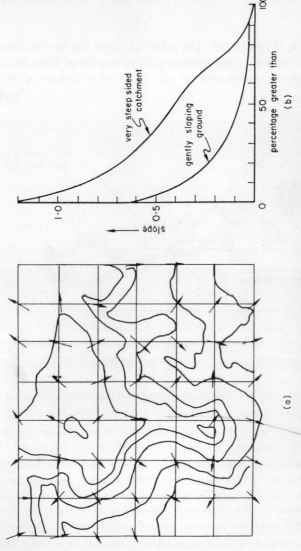

Fig. 6.14 (a) *Rectilinear grid to evaluate catchment slope and orientation.* (b) *Frequency curves for comparison of catchment steepness*

pattern. The grid of Fig. 6.14(a) can be used for this feature also, by the evaluation of the angle between slope direction and the N–S meridian (say) at each grid point and the subsequent plotting of a circular frequency diagram like that of Fig. 6.15, similar to a wind rose.

(d) *Shape of catchment.* The effect of shape can best be demonstrated by considering the hydrographs of discharge from three differently shaped catchments of the same area, shown in Fig. 6.16,

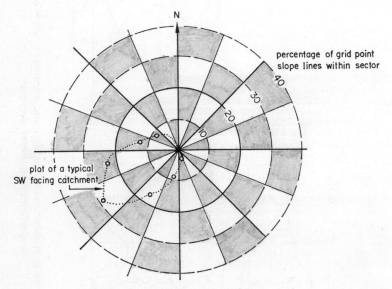

Fig. 6.15 *Orientation diagram*

subjected to rainfall of the same intensity. If each catchment is divided into concentric segments, which may be assumed to have all points within an equal distance along the stream channels from the control point, it may be seen that the shape A will require 10 time units (say hours) to pass before every point on the catchment is contributing to the discharge. Similarly B will require 5 h and C $8\frac{1}{2}$ h. The resulting hydrographs of runoff will be similar to those shown in Fig. 6.16, each marked with the corresponding lower-case letter. B gives a faster stream rise than C and A, and similarly a faster fall, because of the shorter travel times.

This factor of shape also affects the runoff when a rainstorm does not cover the whole catchment at once but moves over it from one end to the other. For example, consider catchment A to be slowly covered by a storm moving upstream which just covers the catchment after 5 h. The last segment's flood contribution will not arrive at the control for 15 h from commencement, so that the hydrograph

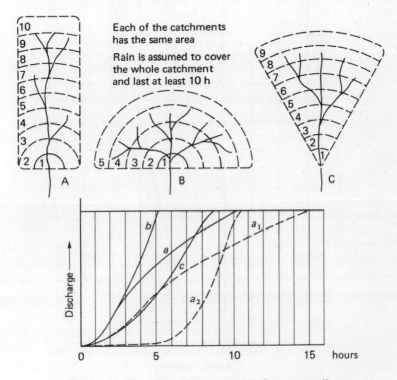

FIG. 6.16 *The effect of shape on catchment runoff*

(a) of Fig. 6.16 will now have the form of curve a_1 on that figure. Alternatively if the storm were moving at the same rate downstream, the flood contribution of time-segment 10 would arrive at the control point only 5 h after that of segment 1, so that a rapid flood rise (a_2 in Fig. 6.16) would occur. The effect of changing the direction of storm movement on the semicircular and fan-shaped catchments will be less marked than this but still appreciable.

(e) *Altitude of the catchment.* Generally, precipitation increases with altitude though individual catchments show wide variations from the general rule. Fig. 6.17 shows the trend for S.W. Scotland. More important however is the effect of reduced evaporation in

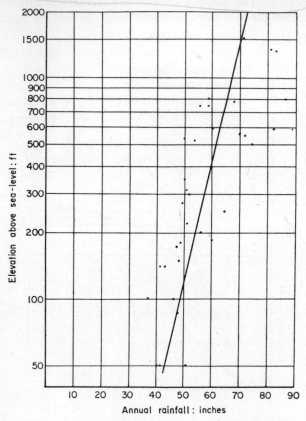

FIG. 6.17 *Correlation between altitude and annual rainfall for S.W. Scotland*

lower temperatures and the temporary storage of precipitation as snow and ice. This affects the distribution of the mean monthly runoff reducing it to a minimum in winter in cold climates. There are thus counterbalancing effects with increasing altitude and which one predominates depends on the prevailing climate. Comparisons of different catchments may be made by plotting frequency distributions of percentage catchment area at or above certain altitudes.

(f) *Stream pattern.* The pattern of stream development in a catchment can have a marked effect on the rate of runoff. A well-drained catchment will have comparatively short times of concentration and hence a steeper flood-rise hydrograph than a catchment with many surface depressions, marshy ground and minor lakes for example. It is very difficult to convert particular drainage patterns to numbers, but comparisons may be made in terms of *stream density,* i.e. length of stream channel per unit area; average length of tributaries or average length of overland flow, i.e. flow to the nearest stream channel. Geology of course plays a part in the process of stream formation, and both geological and topographical maps should be examined in assessing the effect of different patterns.

(g) *Other factors.* In addition to the main factors discussed above, the following will influence the rate and total quantity of surface runoff and so affect the magnitude of flood peaks and hydrograph shape.

(i) condition of stream channels: whether clean or weedy and overgrown.
(ii) the presence of reservoirs, lakes, flood plains, swamps etc. (see (f) above).
(iii) land use: e.g. whether arable land, grassland or forest: artificial drainage etc.
(iv) Sub-surface conditions including initial soil-moisture, height of the phreatic surface, depth and permeability of aquifers and infiltration capacities.

6.5 Climatic factors

In section 6.4(d) reference has already been made to the effect of storm movement on surface runoff. If the areal extent of a storm is such that it does not cover the whole of a catchment, the runoff will be less than from complete coverage.

In section 6.4(e) the effect of the form of precipitation was mentioned, where snowfall and freezing temperatures can effectively put the expected runoff into storage and reduce evapo-transpiration.

The main effect of climate however is in rainfall intensity and *duration.* Rainfall intensity has a direct bearing on runoff since once the infiltration capacity is exceeded all the excess rain is available and flows to the surface water-courses. Intensities vary greatly, the maximum usually occurring in severe local thunderstorms. It will be realised that for severe local maxima to be recorded at all can only occur fortuitously, so that the highest recorded intensities will certainly have been exceeded many times. Jennings [42] and subsequently Paulhus [87] list the highest recorded values, some of which are as follows:

Duration	Depth		Station	Date
	in.	mm		
1 min	1·23	31	Unionville Md., U.S.A.	4 July 1956
20 min	8·10	206	Curtea de Arges, Rumania	7 July 1889
24 h	73·62	1870	Cilaos, La Reunion	15–16 Mar 1952
31 days	366·14	9300	Cherrapunji, India	July 1861

The highest intensities recorded in the British Isles are given below:

Duration	Depth		Station	Date
	in.	mm		
10 min	1·30	33	Chagford	23 Sept. 1927
about 60 min	3·63	92	Maidenhead	12 July 1901
105 min	6·09	155	Hewenden Reservoir	11 June 1956
about 14½ h	11·00*	279	Martinstown, Dorset	18 July 1955

* The storm of Fig. 2.12

Since intensity represents depth/time, it cannot be considered separately from duration. The same depth of rainfall delivered over two different durations will obviously produce quite different runoff rates. What can be said is that different climates will produce different meteorological conditions leading to different types of rain which may inherently have quite dissimilar durations. For example, in England a thunderstorm may conceivably produce rainfall intensities as great as 20 mm/min but it cannot be imagined that such storms can last for more than a period of minutes, whereas the *monsoon* rain in India can fall continuously for weeks at average intensities greater than 10 mm/h, a condition never approached in most other parts of the world. (See also Section 9.6.)

The influence of duration on the hydrograph of runoff may be seen from Fig. 6.18, where a *uniform-intensity storm* causes the hydrograph of stream-rise *a*. Such storms may be defined as covering the whole catchment area, over which the depth of rainfall is reasonably constant and delivered at a constant rate. Although rare in nature, they are used in hydrology to determine characteristics of catchments. After a certain time, t_c, the period of concentration, the rate of runoff becomes constant. A hydrograph of this form is typical only of very small catchments; e.g. paved urban areas, where such constant runoff is quickly achieved. Natural catchments of any size

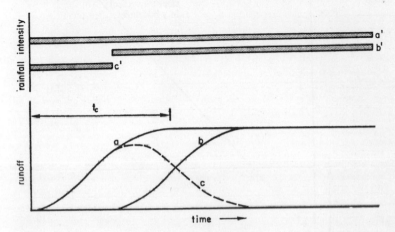

FIG. 6.18 *Hydrograph of short-period storm runoff derived from two long-duration uniform-intensity storms*

have periods of concentration longer, as a general rule, than periods of uniform-intensity rain. The effect of short periods of rain can be found by subtracting the hydrographs of two long periods, exceeding t_c, and separated in time by the short period, one from the other. In Fig. 6.18, *a* and *b* are similar and are the result of rainfalls *a'* and *b'* respectively. Their subtraction leaves the short-period rain *c'* and its resulting hydrograph *c*, which is the typical shape of most natural hydrographs.

6.6 Rainfall/runoff correlation

While there is a general cause-and-effect relationship between rainfall and the resulting runoff, it will be clear by now that it is not a direct one. By the time that evaporation, interception, depression storage, infiltration and soil-moisture deficiency are taken into account and the resulting residual rainfall at various intensities applied to catchments that vary in size, slope, shape, altitude, sub-surface geology and climate, the relationship must include extreme values that defy rational correlation, at least in the short term.

Notwithstanding the foregoing, it may be possible to establish an empirical relationship for a particular catchment based on annual

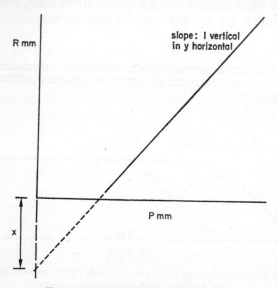

Fig. 6.19 *Rainfall/runoff correlation*

precipitation and runoff. To do this it is best to use a *water-year* rather than a calendar year. This is a 12-month period starting and finishing at the time of seasonal minimum flow. If precipitation and runoff are plotted against each other as catchment depths, a correlation like that shown in Fig. 6.19 may be obtainable. In temperate and tropic, humid climates such a straightline relationship generally is found, where if P is annual precipitation, annual runoff R is expressed as

$$R = \frac{P}{y} - x \qquad (6.7)$$

so that the annual rainfall can be used to obtain a first approximation to the annual runoff.

Variations from the straight line may be due to conditions in the preceding year which gave markedly higher or lower groundwater levels, or to variations in the seasonal distribution of rainfall. The method can be used also for wet months in humid climates when the ground is saturated, but beyond such narrow limits it is not valid.

Although the application of a relationship like Eq. (6.7) is restricted, it can nevertheless be a useful method for estimating total annual runoff of completely ungauged catchments, if they are in similar climates and of similar size and character.

Further refinements are possible taking into account the particular period of the year, the antecedent precipitation index (see Section 4.4.2) and the storm duration as well as depth, so that relationships may be derived for particular storms on a particular catchment. Coaxial graphs may be produced which take the various variables into account. A relationship of this form is shown in Fig. 4.7.

7 Hydrograph Analysis

7.1 Components of a natural hydrograph

The various contributing components of a natural hydrograph are shown in Fig. 7.1. To begin with there is *base-flow* only, i.e. the groundwater contribution from the aquifers bordering the river which go on discharging more and more slowly with time. The hydrograph of base-flow is near to an exponential curve and the quantity at any time may be represented very nearly by

$$Q_t = Q_o e^{-\alpha t}$$

where Q_o = discharge at start of period

Q_t = discharge at end of time t

α = coefficient of aquifer

e = base of natural logarithms

As soon as rainfall begins there is an initial period of interception and infiltration before any measurable runoff reaches the stream channels and during the period of rain these losses continue in a reduced form as discussed previously, so that the *rain graph* has to be adjusted to show *nett*, or *effective rain*. When the initial losses are met, surface runoff begins and continues to a peak value which occurs at a time t_p, measured from the centre of gravity of the rain-graph of nett rain. Thereafter it declines along the *recession limb* until it completely disappears. Meantime the infiltration and percolation which has been continuing during the gross rain period results in an elevated groundwater table which therefore contributes more at the end of the storm flow than at the beginning, but thereafter is again declining along its *depletion curve*.

Surface runoff is, for convenience, assumed to contain two other components; *channel precipitation* and *interflow*. Channel precipitation is that portion of the total catchment precipitation that falls directly on the stream, river and lake surfaces. It is usually small but if large lakes are present in the catchment it may be quite important and then requires separate treatment. Interflow refers to water

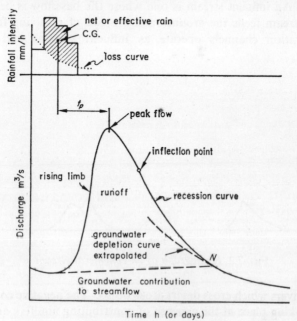

FIG. 7.1 *Component parts of a natural hydrograph*

travelling horizontally through the upper horizons of the soil, perhaps in artificial tile drain systems or above *hard-pans* or impermeable layers immediately below the surface. Such flow can vary from nothing to appreciable fractions of total runoff.

Since the groundwater contribution to flood flow is quite different in character from surface runoff it should be analysed separately and one of the first requirements in hydrograph analysis therefore is to separate these two.

7.2 The contribution of baseflow to stream discharge

Since baseflow represents the discharge of aquifers, changes occur slowly and there is a lag between cause and effect which may easily extend to periods of days or weeks. This will depend on the *trans-missibility* of the aquifers bordering the stream and the climate. Some of the infinite number of natural conditions are considered below.

A broad distinction should be made between *influent* and *effluent* streams. An influent stream is one where the baseflow is negative, i.e. the stream feeds the groundwater instead of receiving from it, e.g. irrigation channels operate as influent streams and many

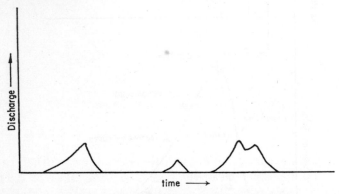

FIG. 7.2 *Hydrograph of an ephemeral stream*

natural rivers which cross desert areas do so. The negative contribution is taking place at the expense of contributing aquifers on other parts of the stream, since there can be no baseflow from a *wholly* influent stream. Such a stream, e.g. a Middle Eastern *wadi*, will dry up completely in rainless periods and is called *ephemeral*, and has a hydrograph of the form of Fig. 7.2.

An effluent stream on the other hand is fed by the groundwater and acts as a drain for bordering aquifers. The great majority of streams in Britain and Europe are in this category.

Intermittent streams are those which act as both influent and effluent streams according to season, tending to dry up in the dry season.

Perennial streams are greatly in the majority, with a low dry-season flow fed by baseflow and are mainly effluent streams, though many perennial rivers crossing different geological formations of varying permeability and subject to different climates are both influent and effluent at different parts of their courses. A good example of this is the river Euphrates in Iraq. Fig. 6.9 shows a part-annual hydrograph of the Euphrates and the slow seasonal variation of the baseflow may be observed. This baseflow is derived principally from the headwaters of the catchment in northern Iraq, Turkey and Syria. At Hit, where the hydrograph was observed, the river for much of the year is influent.

Bank storage describes the portion of runoff in a rising flood which is absorbed by the permeable boundaries of a water course above the

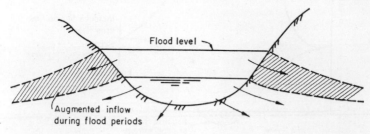

Fig. 7.3 *Influent stream*

normal phreatic surface. It is illustrated in Figs. 7.3 and 7.4. In the latter figure the direction of the arrows showing influx of groundwater to the stream will be reversed during the flood period while

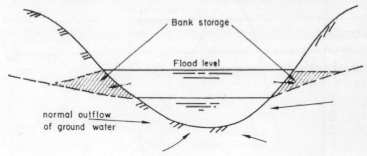

Fig. 7.4 *Effluent stream*

the surface level of the stream is above the phreatic surface. As a result the hydrograph of a particular flood might well have a baseflow contribution as indicated in Fig. 7.5. Such a separation as is shown there is virtually impossible to make quantitatively but it is qualitatively correct.

In many natural rivers, depending naturally on bank permeability and the slope of the phreatic surface, the variation in baseflow will

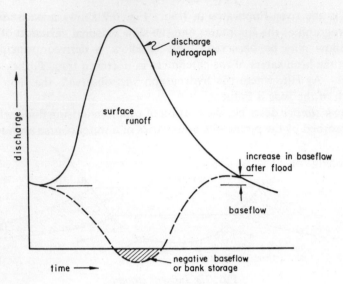

FIG. 7.5 *Negative baseflow*

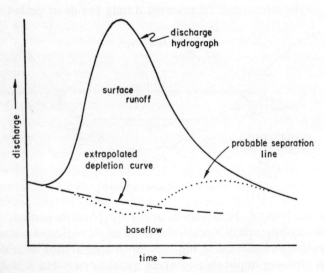

FIG. 7.6 *Baseflow separation*

124

be much less than indicated in Fig. 7.5 and will only cause a slight dip from the extrapolation of the depletion curve, followed by a gradual rise to a higher-than-initial value as indicated in Fig. 7.6.

7.3 Separation of baseflow and runoff

It has been shown in the preceding paragraphs that the dividing line between runoff and baseflow is indeterminate and can vary widely. Since, to *analyse* its precise position would require a detailed knowledge of the geohydrology of the catchment, including the areal extent and transmissibility of the aquifers, it is generally more practical to use a consistent separation technique. Which of the following is used depends on the data available.

If a continuous discharge record of the stream, over a period of a few years is available, the hydrograph should be plotted in the manner of Fig. 7.7(a). This is examined for portions which include recession curves running into baseflow contribution only, after runoff has ceased, at as many different stages as possible. These sections are abstracted from the continuous hydrograph and plotted again to a log Q vertical scale and linear time scale, as shown in Fig. 7.7(b). Starting with the lowest recession flow in the record, a curve is now constructed such as is tangential to the lower portions (i.e. the true depletion curves) of the log Q abstracted plots. This is most easily done by moving tracing paper over the plots, with the abscissae coincident, until each log Q plot in successive increasing magnitude fits into the growing curve and extends it fractionally upward. The tangential curve thus established to the highest stage possible is then converted back to linear vertical scale and is called the *master depletion curve* for the particular gauging station. It may now be applied to the hydrograph of a particular storm period in the manner indicated in Fig. 7.8 whereby the depletion curves are fitted together at their lower ends and the point of divergence marked as N. N represents the point at which surface runoff has effectively ceased and a straight line is drawn to it from the point of sudden rise. This line, shown dashed in Fig. 7.8, represents the base line of the hydrograph of surface runoff which can then be analysed.

While the procedure outlined above is probably the best available, it does depend on previously observed data which is not always available. An alternative procedure is to establish the point of greatest curvature on the recession limb of the hydrograph. This is

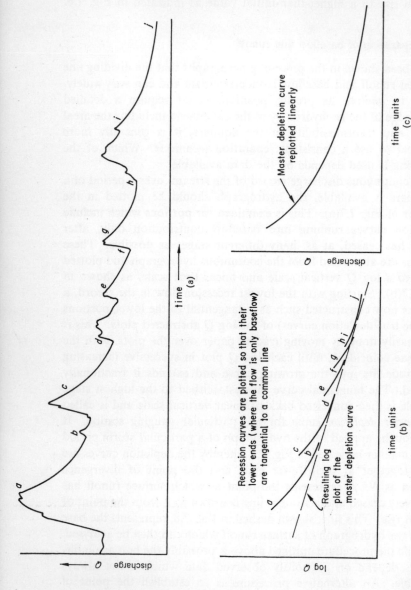

FIG. 7.7 *Derivation of a master depletion curve. (a) Normal hydrograph with recession curves selected. (b) Log plot of recession curves. (c) Linear plot of master depletion curve*

In the figure labels:

(a) discharge Q — time

(b) $\log Q$ — Recession curves are plotted so that their lower ends (where the flow is only baseflow) are tangential to a common line. Resulting log plot of the master depletion curve. — time units

(c) discharge Q — Master depletion curve replotted linearly — time units

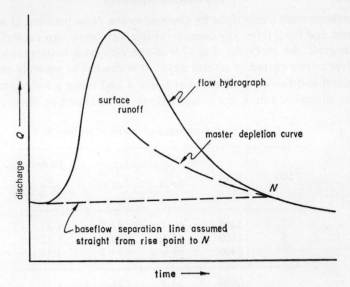

FIG. 7.8 *Procedure to separate base flow*

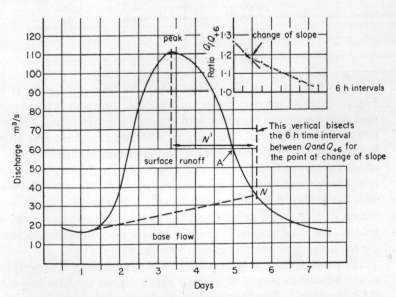

FIG. 7.9 *Alternative method of separating baseflow*

perhaps most easily done by computing the ratio between Q at any time and say 2 h (or any convenient interval) later. An example will illustrate the method. Fig. 7.9 is the observed hydrograph of a river over a period of several days. It is desired to separate surface runoff and baseflow. Starting at point A and using a 6 h separation for successive ratios, the computations are as shown in Table 7.1.

TABLE 7.1. *Computation to assist in finding N*

Day	Hour	Q m³/s	Q_{+6} m³/s	Ratio $\dfrac{Q}{Q_{+6}}$
5	1200	60·1	47·5	1·27
	1800	47·5	39·0	1·22
	2400	39·0	33·2	1·18
6	0600	33·2	28·6	1·16
	1200	28·6	25·2	1·13
	1800	25·2	22·7	1·11
	2400	22·7	20·9	1·09
7	0600	20·9	19·7	1·06
	1200	19·7	18·9	1·04
	1800	18·9	18·2	1·04
	2400	18·2	—	

It may be seen from the inset graph of the ratio–time interval on Fig. 7.9 that two separate slopes are apparent, the upper being associated with runoff and the other with groundwater depletion. At their intersection, the critical ratio may be determined and the first point beyond the region of intersection on the groundwater side gives a conservative position for N. The subsequent hydrograph analysis is not very sensitive to the precise position of N and either 0300 or 0600 h on Day 6 would be satisfactory. A straight line is now drawn to N from the point where the hydrograph started to rise, as before. The total amount of runoff may now be obtained by measuring the area under the curve and above the straight line.

The position of N is important also in synthesising hydrographs, as will be seen in Sect. 7.10, since it partly defines the *base length* of the hydrograph. The baselength is made up of the part before the peak, which depends on the length of the period of rain and t_c the period of concentration, and the recession limb after the peak which depends primarily on the character of the catchment. From observations on many natural catchments the position of N may be established empirically, from the Table 7.2 given below, in days after the peak of the flood.

TABLE 7.2. *Catchment area as a guide to N*

Catchment Area km²	Time from peak to N days
250	2
1250	3
5000	4
12500	5
25000	6

7.4 The unit hydrograph

Having derived the hydrograph of surface runoff by the methods discussed in preceding sections, the problem now arises of how it can be correlated with the rainfall which caused it. Clearly the quantity and intensity of the rain both have a direct effect on the hydrograph but it has not yet been made clear how, and to what extent, each of these affects it. The method of doing this is a part-empirical, part-theoretical technique which uses the concept of the *unit hydrograph* (also called the *unitgraph*), first described by Sherman [43].

It should be emphasised that the correlation sought is between the *nett* or *effective rain* (i.e. the rain remaining as runoff after all losses by evaporation, interception and infiltration have been allowed for) and the surface runoff (i.e. the hydrograph of runoff minus baseflow).

The method involves three principles which are as follows:

1. With uniform-intensity nett rainfall on a particular catchment, different intensities of rain of the same duration produce runoff for the same period of time, although of different quantities. This is an empirical rule which is approximately true and is illustrated in Fig. 7.10.
2. With uniform-intensity nett rain on a particular catchment, different intensities of rain of the same duration produce

hydrographs of runoff, the ordinates of which, at any given time, are in the same proportion to each other as the rainfall intensities. That is to say, that n times as much rain in a given time will give a hydrograph with ordinates n times as large. In Fig. 7.10 the ordinates at time t_1 are np and p respectively for rainfall intensities of ni and i.

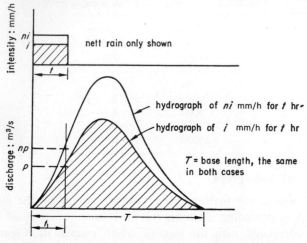

FIG. 7.10 *Proportional principle of the unitgraph*

3. The principle of superposition applies to hydrographs resulting from contiguous and/or isolated periods of uniform-intensity nett rain. This is illustrated in Fig. 7.11 where it may be seen that the total hydrograph of runoff due to the three separate storms is the sum of three separate hydrographs.

Having established these principles the concept of unit rain is now introduced. A unit of rain may be any specified amount, measured as depth on the catchment, usually 1 cm or 1 in. but not exclusively so. The unit rain then must all appear as runoff in the *unit hydrograph*. The area under the curve of the hydrograph has the dimensions of instantaneous discharge multiplied by time, or

$$\frac{L^3}{T} \times T = L^3 = \text{volume of runoff}$$

so that although unit rain is spoken of as 1 cm over the whole of the catchment area the resulting runoff is given in cubic metres, and the quantities involved are identical. If the unit graph for a particular catchment, *and a particular duration of rain* is known, then from principle 2, the runoff from any other rain, of the same duration may be predicted.

This is a first step towards the complete correlation sought, but if the rainfall should be of different duration from that of the unit-graph then the unitgraph must be altered before it can be used.

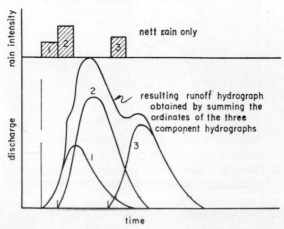

FIG. 7.11 *Principle of superposition applied to unitgraphs*

7.5 Unit hydrographs of various durations

7.5.1 Changing a short duration unitgraph to a longer duration unit-graph. The simplest way to produce a unitgraph for a longer duration of rain is illustrated in Fig. 7.12.

Suppose a 2 h unitgraph is given and a 4 h unitgraph is wanted. This may be obtained by assuming a further 2 h period of nett rain immediately following the first, which will give rise to an identical unitgraph, but shifted to the right in time by 2 h. If the two 2 h unitgraphs are now added graphically, the total hydrograph obtained represents the runoff from 4 h of rain at an intensity of ½ cm/h. (This must be so because the 2 h unitgraph contains 1 cm rain.) This total hydrograph is therefore the result of rain at twice the intensity required and so the 4 h unitgraph is derived by dividing its ordinates by 2. This is shown as the dashed line on Fig. 7.12. It will be observed that it has a longer time base by 2 h than the 2 h

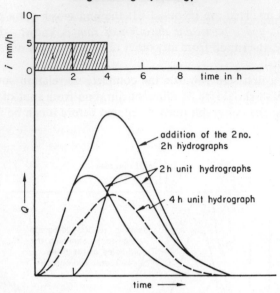

FIG. 7.12 *Changing a short period unitgraph to a long period one*
(if the long is an even multiple of the short)

unitgraph; this is reasonable since the rain has fallen at a lower
intensity for a longer time.

7.5.2 Changing a long duration unitgraph to a shorter duration unit-graph.

To derive a short-rain period unitgraph from that for a
longer period it is necessary to use an S-curve technique. An S-
curve is simply the total hydrograph resulting from a series of
continuous uniform-intensity storms delivering 1 cm in t_1 h on the
catchment, i.e. it is the hydrograph of runoff of continuous rainfall
at an intensity of $1/t_1$. Such a hydrograph has the form of Fig. 7.13,
the discharge of the catchment becoming constant after t_c, the time
of concentration, when every part of the catchment is contributing
and conditions are in a steady state. Thus each S-curve is unique
for a particular unitgraph duration, in a particular drainage
basin.

If a second S-curve is drawn one unit period to the right of
the first, then clearly the difference between the two S-curves expressed graphically equals the runoff of one t_1 h unitgraph.

If the unitgraph for a shorter period storm of t_2 h is required, it may be obtained by drawing the S-curve again, but shifted only t_2 h along the time axis. The graphical difference between ordinates of the two S-curves now represents the runoff of t_2 h rain at an intensity of $1/t_1$ cm/h. The ordinates of this S-curve difference graph must therefore be multiplied by t_1/t_2 so that the rain intensity represented is $1/t_2$ cm/h, which is the intensity required for the t_2 unitgraph. The procedure is illustrated in Fig. 7.13.

If the time base of the unitgraph is T h, then steady state runoff must occur at T h and so only T/t_1 unitgraphs are necessary to develop constant outflow and so produce an S-curve. The constant

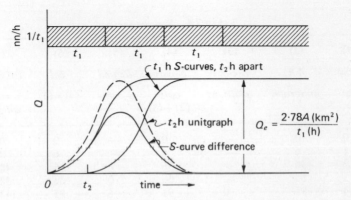

FIG. 7.13 *Transposing unitgraphs by S-curves*

outflow, Q_e, may easily be obtained since 1 cm on the catchment is being supplied and removed every t_1 h.

$$Q_e = \frac{2 \cdot 78 A}{t_1}$$

where A is catchment area in km²
$\quad\quad t_1$ is duration in h
and $\quad Q_e$ is in m³/s.

It will be apparent that the method may be used for altering the unit period either way, longer or shorter and that if changing from shorter to longer duration, that t_2 need not be a direct multiple of t_1. Although the method has been described graphically, in practice its application is usually made in tabular form and Example 7.1 illustrates it.

Example 7.1. *Given the* 4 *h unit hydrograph listed in column* (2) *derive the* 3 *h unit hydrograph. The catchment area is* 300 *km²*.

The S-curve equilibrium flow $Q_e = \dfrac{2 \cdot 78 \times 300}{4} = 208 \ m^3/s$.

It will be noted that $Q_e = 208 \ m^3/s$, as calculated, agrees very well with the tabulated *S*-curve terminal value 207. This is an indication that the 4 h period of the unit graph is correctly assessed. Very often with an uneven rainfall distribution, an attempt has to be made to reduce the nett rain to a uniform-intensity rain of particular duration. The *S*-curve can in this way serve as a check on the chosen

TABLE 7.3. *S-curve method*

(1) time h	(2) 4 h unitgraph	(3) S-curve additions	(4) S-curve cols (2)+(3)	(5) lagged S-curve	(6) col (4)−(5)	(7) col 6 × $\frac{4}{3}$ = 3 h unitgraph
0	0	—	0	—	0	0
1	6	—	6	—	6	8
2	36	—	36	—	36	48
3	66	—	66	0	66	88
4	91	0	91	6	85	113
5	106	6	112	36	76	101
6	93	36	129	66	63	84
7	79	66	145	91	54	72
8	68	91	159	112	47	63
9	58	112	170	129	41	55
10	49	129	178	145	33	44
11	41	145	186	159	27	36
12	34	159	193	170	23	31
13	27	170	197	178	19	25
14	23	178	201	186	15	20
15	17	186	203	193	10	13·5*
16	13	193	206	197	9	12*
17	9	197	206	201	5	6·5*
18	6	201	207	203	4	5·5*
19	3	203	206	206	0	0*
20	1·5	206	207	206	1	1·5*
21	0	206	206	207	−1	

* Slight adjustment is required to the tail of the 3 h unitgraph. This is most easily done by eye (see Fig. 7.14). All figures except col. 1 in m³/s value. If the S-curve terminal value had fluctuated wildly and not steadied to a minor variation it would have indicated an incorrect rainfall-time for the unit graph.

Note also that it was not necessary in Table 7.3 to set out T/t_1 columns of the 4 h unitgraphs, and add them laterally. The *S-curve additions* are the S-curve ordinates shifted in time by 4 h. Since the first 4 h of unitgraph and S-curve are the same, the S-curve additions and S-curve columns are filled in, in alternate steps. The effect is the

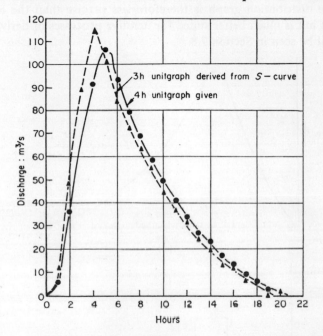

FIG. 7.14 *Unitgraph derived by S-curve method*

same as setting out rows of unitgraph ordinates successively staggered 4 h, since the S-curve additions represent the sum of all previous unitgraph ordinates.

7.6 The unit hydrograph as a percentage distribution

The distribution graph, first used by Bernard [44], represents the unitgraph in the form of percentages of total flow occurring in particular unit periods. Since the discharge represented by a unitgraph is directly proportional to nett rain, the percentages in unit times will remain constant whatever the nett rain. This is a useful means of applying the unitgraph method in some cases.

In Fig. 7.15 a unit hydrograph is shown, together with the derived distribution graph which represents it. The areas under the curve and under the step line are the same and so, in deriving unitgraphs from distribution percentages a smooth line must be drawn through the steps to give equal areas.

The distribution graph is therefore less precise than the hydrograph but is much better suited for iterative processes of derivation, as will be seen in Section 7.8.

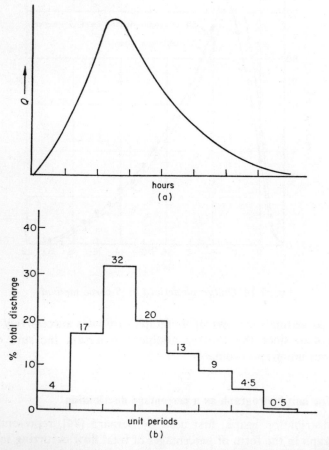

FIG. 7.15. (a) *Unit hydrograph.* (b) *Derived distribution graph*

7.7 Derivation of the unit hydrograph

The unitgraph for a particular catchment may be derived from the natural hydrograph resulting from any storm which covers the catchment and is of reasonably uniform intensity. If the catchment is very large, i.e. greater than (say) 5000 km², it may never be covered by a uniform-intensity storm, since these are limited in size by meteorological conditions. In such a case the catchment should be divided

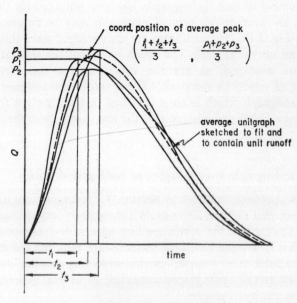

FIG. 7.16 *Average unit hydrograph from a number*

up into tributary catchments and the unit graphs for each of these determined separately.

The first step is to separate the baseflow from surface runoff (Sect. 7.3) and plot the runoff and the rain graph on the same time base. The quantity of nett storm rain must then be estimated and its intensity and duration established. A check is now made on the quantity of nett rain on the catchment and the amount of runoff under the hydrograph. These should be the same and one or the other may require adjustment.

The unitgraph may now be obtained by dividing the runoff

hydrograph ordinates by the nett rain in cm. The adjusted ordinates represent the unit graph for the particular duration established.

It is always advisable to determine several unit graphs, using separate and distinct isolated uniform intensity storms, if available. Natural events like rainstorms and runoff, are affected by a multiplicity of factors and no two are precisely the same. Frequently the best natural data will be for different rain durations and the resulting unit graphs will require to be altered to the same duration (Sect. 7.5). Once a number of such hydrographs has been obtained for the same duration, an 'average', or typical unitgraph may be determined as shown in Fig. 7.16. The ordinates are *not* averaged since this would produce an untypical peak. The peak values of the separate unitgraphs are averaged, as are the values of the time from the beginning of runoff to the peak. These values are assigned to the average unitgraph which is then sketched in to a median form on both rising and falling limbs, so that the total area under the curve is equal to 1 cm runoff.

7.8 Unit hydrographs from complex or multi-period storms

While the approach outlined in Section 7.7 is simple and direct, it presupposes that the records contain a number of isolated, uniform-intensity storms and the corresponding natural hydrographs. Frequently this is not the case and methods are required for deriving unitgraphs from more complex storms, involving varying intensities of rain with runoff hydrographs consisting of several superimposed separate storm hydrographs.

To derive unit graphs from such records is more laborious than for simple storms but a variety of methods is available, two of which are discussed below.

The first, described by Linsley, Kohler and Paulhus [45] requires the writing and successive solving of a series of equations for each ordinate of the complex hydrograph, baseflow being assumed previously separated. The process may be illustrated by reference to Fig. 7.17.

The first rain period, of duration and intensity t and i_1 respectively, gives rise to runoff illustrated by the hypothetical hydrograph bounded by the lower dashed line. Each ordinate of this hydrograph is ti_1 times the unitgraph ordinate $U_1, U_2 \ldots U_n$. Similarly, the second and third rains of intensity i_2 and i_3 respectively produce additional runoff whose ordinates in each case are ti_2 and ti_3, multiples of the t h unitgraph shifted appropriately in time. If the complex hydrograph is now defined by ordinates at suitable equal intervals (conveniently but not essentially fixed as a whole multiple of t h) then the first ordinate of the unitgraph, U_1, may be obtained from $Q_1 = ti_1U_1$ where Q_1 is the observed runoff, hence U_1 may be found.

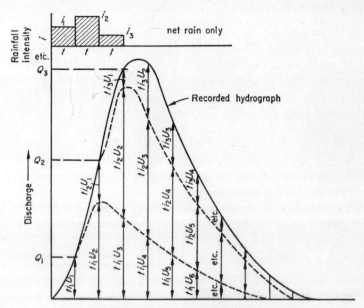

FIG. 7.17 *Derivation of unit hydrograph from a multi-period storm hydrograph*

For the second ordinate, $Q_2 = ti_1U_2 + ti_2U_1$ in which equation U_2 is the only unknown.

The third ordinate is similarly obtained from $Q_3 = ti_1U_3 + ti_2U_2 + ti_3U_1$ where U_3 is now the only unknown. Proceeding in this way, the t h unitgraph ordinates may be successively determined.

In the above illustration all the rain periods, although of different intensities, were assumed of the same duration t h. This is a condition for the use of the method since otherwise other variables U_1', U_2' etc. (the ordinates of the t' h unitgraph) would be introduced.

Although the method appears simple, since each U ordinate depends on the preceding ones, and on the assumptions about intensity and duration of rainfall and deduction of an assumed baseflow, errors accumulate, and several trials and restarts may be necessary to find a reasonable unitgraph.

The second method is due to Collins [46] and is amongst the simplest of various iterative methods proposed. To illustrate its use a unit hydrograph will be derived from the data of rainfall and natural discharge for the catchment of the River Rother at Woodhouse Mill in Yorkshire.

Collin's method for the determination of a unit hydrograph from a multi-period storm. This method requires the initial selection of a set of coefficients, or distribution percentages (see Sect. 7.6) of the unitgraph. This distribution graph is then applied to the various rain periods, excepting the largest one, and the resulting discharge subtracted from the actual discharge to obtain a set of 'residuals'. These residuals should represent the discharge of the unitgraph applied to the largest rain. If the correspondence is poor, the initial coefficients are altered and another trial is made. By a series of converging approximations, the residual graph is made to correspond with the assumed distribution graph.

The procedure is set out below step by step and referred to the particular case of the River Rother at Woodhouse Mill, for which a unit hydrograph is derived from the storm of May 14/15, 1967. Phase A is standard procedure; Phase B is the Collins approach.

Phase A: Assembling and preparing the data:

1. Assemble all rainfall data available for the catchment under consideration and the storm period, including daily observations, recording rain gauge records and the synoptic weather maps of the region, if available.
2. Derive a mean mass curve of rainfall for the catchment, for the period of rain producing the hydrograph under study. Make provisional separation of the rain into separate uniform periods.
   The catchment of the River Rother at Woodhouse Mill, Yorkshire is shown in Fig. 7.18. The continuous rainfall record (mass curve) at Sutton-in-Ashfield for the period of the storm is shown in Fig. 7.19. The dashed line superimposed on the mass curve represents the idealised intensities used and plotted in the appropriate time period of Fig. 7.20. The total rainfall depth is measured daily at Chesterfield in the centre of the catchment.

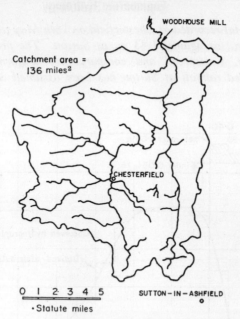

Catchment area = 136 miles²

0 1 2 3 4 5
· Statute miles

WOODHOUSE MILL

CHESTERFIELD

SUTTON - IN - ASHFIELD

FIG. 7.18 *Catchment of the River Rother at Woodhouse Mill*

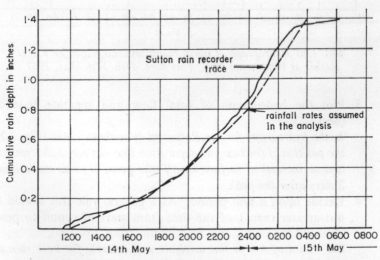

FIG. 7.19 *Rainstorm of 14th–15th May, 1967, at Sutton-in-Ashfield*

The total recorded at Chesterfield on 15th May at 9.00 a.m. was
1·42 in., as against 1·43 in. at Sutton. The precipitation was
frontal, widespread and comparatively uniform and so the
recorded rainfall at Sutton has been assumed catchment-wide.

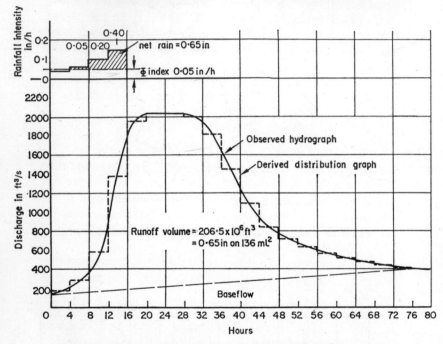

FIG. 7.20 *Hydrograph of storm runoff and raingraph for River*
Rother at Woodhouse Mill, Yorkshire, 14th–17th May, 1967

3. Plot the hydrograph of total flow and separate baseflow
   (Sect. 7.3.).
   In the hydrograph for the Rother at Woodhouse Mill (Fig. 7.20)
   the position of the baseflow separation line was very indeterminate
   and so the end of surface runoff (the point N) was made about
   2 days after the peak.
4. Decide upon a unit period. As a general rule this should be
   not greater than $\frac{1}{4}$ of the time from start of runoff to peak.
   Consider the provisional rain graph (from 2 p. 140).
   The choice of 4 h in this analysis was suitable for both rain and
   runoff.

5. Consider the soil-moisture deficit of the catchment and use Φ index or f_{av} method for estimating storm loss for each rain period. If antecedent precipitation indices have been kept for the catchment, use these. Make an estimate of initial and subsequent loss rates. Compare the nett rain derived in this way with the surface runoff expressed as depth on the catchment. If the two do not agree, one or the other must be adjusted. *The Rother catchment was wet, but not saturated, from light rains the day previous to the storm. It seemed reasonable to assume the first period's rain lost entirely in making good the remaining soil-moisture deficit. The index was subsequently chosen to balance the nett rain with surface runoff.*

Phase B: Using the data to derive the unitgraph:

6. Tabulate the relevant data in columns 1–9 of a table similar to Table 7.4 and provide columns for the number of unit periods in the unit hydrograph under 'Distribution coefficients'.
   The number of columns in Table 7.4 under 'Distribution coefficients' is 16. This is the number of unit periods from the beginning of the last rain period to the end of surface runoff.
7. Assume distribution coefficients of the unitgraph (representing percentage of total runoff per unit period) and arrange in appropriate columns.
8. Determine the discharge, which, flowing constantly for one unit period, would just equal 1 in. nett rain on the catchment. *This figure is found in this case to be:*

$$\frac{136 \times 27.9 \times 10^6}{12 \times 4 \times 3600} = 21960 \text{ ft}^3/\text{s}$$

9. The first nett rain is multiplied by this discharge and the product is distributed in percentages across the distribution coefficient columns by being multiplied in turn by each percentage coefficient. The various numbers are entered in the columns diagonally.
   In this case $0.05 \times 21960 = 1098$, so in the first column $0.05 \times 1098 = 55$ and so on. Note that 55 was entered opposite the corresponding rain and not on the top row which is ignored.
10. The procedure of 9, above, is repeated for all nett rains except the largest for which a dash is entered throughout.
    It is purely fortuitous that the largest rain is at the end in this case.

TABLE 7.4. Derivation of unit hydrograph from a multi-period storm

(1) Day	(2) Time	(3) Period no.	(4) Rain in.	(5) Losses in.	(6) Nett rain in.	(7) Average	(8) Base-flow	(9) Nett Q	5·0	10·0	14·0	15·0	13·5	11·0	8·5	6·2	4·0	3·3	2·6	2·1	1·7	1·4	1·0	0·7	(11) Σ	(12) Residuals	(13) %
14 May	1200		0·16	0·20	0	180	145	35																			
		1	0·25	0·20	0·05	290	160	130	55																55	75	(0·8)
		2	0·40	0·20	0·20	580	175	405	220	110															330	75	(0·8)
	—2400																										
15 May		3	0·60	0·20	0·40	1380	185	1195		440	154														594	601	6·9
		4				1950	200	1750			615	165													780	970	11·1
		5				2030	215	1815				660	148												808	1007	11·5
		6				2030	225	1805					594	121											715	1090	12·5
		7				1995	240	1755						484	93										577	1178	13·4
		8				1820	255	1565							374	68									442	1123	12·8
	—2400																										
16 May		9				1450	270	1180								273	44								317	863	9·8
		10				1095	280	815									176	36							212	603	6·9
		11				840	300	540										145	29						174	366	4·2
		12				720	315	405											114	24					138	267	3·1
		13				640	325	315												97	19				116	199	2·3
		14				570	340	230													75	15			90	140	1·6
	—2400																										
17 May		15				520	350	170														62	11		73	97	1·1
		16				480	365	115															44	8	52	63	0·7
		17				440	380	60																31	31	29	0·3
		18				420	395	25																	0	25	0·3
2nd trial coefficients									6·4	10·9	12·6	13·5	13·4	12·5	9·4	6·7	4·1	3·2	2·4	1·8	1·3	1·0	0·6	0·4			
3rd trial									7·2	11·5	12·3	12·8	12·9	12·7	9·5	6·7	4·2	3·2	2·5	1·8	1·3	1·0	0·5	0·3			
Accepted coefficients									7·0	11·6	12·5	12·9	12·7	12·0	9·6	6·8	4·3	3·2	2·5	1·8	1·3	1·0	0·5	0·3			100·4

(10) Distribution Coefficients % (the header row values 5·0 10·0 14·0 15·0 13·5 11·0 8·5 6·2 4·0 3·3 2·6 2·1 1·7 1·4 1·0 0·7 are the 1st-trial coefficients).

11. The various discharges are now summed horizontally and entered in the Σ column.
12. The Σ column discharge totals are now subtracted from column 9 and the remainders entered in the 'Residuals' column. These residuals are then converted into percentages of the unit distribution graph by dividing by the discharge of 8 (p. 143) multiplied by the largest rain, and subsequently multiplying by 100. The sum of the percentage column should be 100. The percentages which cannot have been influenced by the largest rainfall are bracketed and redistributed over the other coefficients so that the total, 100%, remains constant. These percentages represent the distribution that would be necessary for the largest rain to make good the nett Q of column 9. If they are the same as the assumed distribution coefficients then the unit distribution graph has been determined.

The residual 1090 of period 6 for example is converted thus

$$\frac{1090}{0\cdot40 \times 21960} \times 100 = 12\cdot5\%$$

The percentages total comes to 100·4 due to rounding off. The bracketed figures have not been redistributed since the next trial requires fairly substantial changes in any case.

13. If the differences between the trial coefficients and the adjusted (after redistribution) coefficients are large, then a new trial set must be adopted and steps 9–12 repeated, until the differences are sufficiently small to be ignored (say <1%). A weighted average of the previous trial and resulting adjusted coefficients should be used, as follows:

If P = the sum of residuals
Q = the sum of discharges of all the periods during which the largest rainfall would have been contributing
C_1 = the trial coefficient
C_2 = the calculated and adjusted coefficient
C_3 = the proposed new trial coefficient

then $\qquad\qquad C_3 = \dfrac{QC_1 + PC_2}{Q + P}$

Alternatively, the weighting is often simply done in proportion to the total rain depth, the trial coefficients having the weighting of rain depths actually used, and the computed coefficients that of the largest rainfall.

14. It is always wise to plot the distribution graph before deciding on acceptable coefficients. It may be found that small adjustments can help to give a smooth curve for the adopted unitgraph.

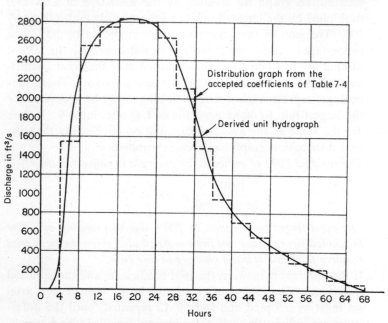

FIG. 7.21 *4-h unit hydrograph for River Rother at Woodhouse Mill derived from the multi-period storm of Fig. 7.19*

In the illustrated case three trials were made with calculations and a final adjustment (without recalculation) made after plotting the distribution graph and hence deriving the unit graph shown in Fig. 7.21.

The method is particularly useful when the largest rainfall is very great compared with the others since rapid convergence of the coefficients then takes place. This was not the case in the example illustrated. Too much reliance should not be placed in unit graphs derived in this way, until they have been used in practice and/or derived from a series of different storms, since the loss rates chosen have a critical influence on the resulting unitgraphs.

7.9 The instantaneous unit hydrograph

An extension of unitgraph theory is the concept of the instantaneous unit hydrograph or *IUH*. The *IUH* is the hydrograph of runoff from the instantaneous application of unit effective rain on a catchment.

Referring to Fig. 7.13 of Section 7.5, the S-curve was seen to be a simple method of deriving a unitgraph of period T h from the unit

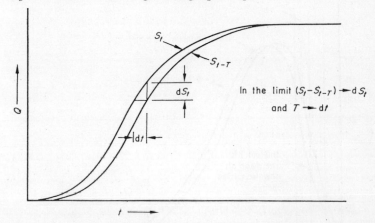

In the limit $(S_t - S_{t-T}) \rightarrow dS_t$

and $T \rightarrow dt$

FIG. 7.22 *The instantaneous unit hydrograph as the S-curve derivative*

graph of any other period t, by drawing two t-h S-curves, T-h apart. This is expressed in the equation

$$U(T, t) = \frac{t}{T}(S_t - S_{t-T}) \qquad (7.1)$$

where $U(T,t)$ represents the ordinates of the T-h unitgraph derived from those of the t-h unitgraph. Now as T, progressively diminishing to zero, $\rightarrow dt$, the right hand side of Eq. (7.1) $\rightarrow$ the S-curve derivation, as may be seen graphically in Fig. 7.22. In equation form this is

$$U(O, t) = \frac{d(S_t)}{dt} \qquad (7.2)$$

i.e. the ordinate of the *IUH* at any time t is given by $\dfrac{dS_t}{dt}$ at time t.

The *IUH* is a unique demonstration of a particular catchment's response to rain, independent of duration, just as the unitgraph is its response to rain of a particular duration. Since it is not time-dependent the *IUH* is thus a graphical expression of the integration of all the catchment parameters of length, shape, slope condition etc. which control such a response.

The conversion of an *IUH* to a unitgraph of finite period is simple. The ordinate of an *n*-h unitgraph at time *t* is the average ordinate of the *IUH* for *n* h before *t*. From Fig. 7.23, it may be seen that the

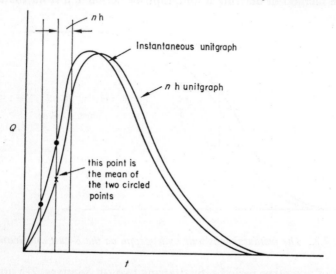

FIG. 7.23 *The n-h unitgraph derived from the IUH*

IUH is divided into *n* h time intervals, and the averages of the ordinates at the beginning and end of each interval are plotted at the end of the intervals to provide the *n*-h unitgraph.

The *IUH* may be used to derive unitgraphs by flood-routing as explained in Section 8.6.

7.10 Synthetic unit hydrographs

In preceding sections it has always been assumed that some records have been available for the derivation of the unitgraph, but there are many catchments for which there are no runoff records at all and for which unitgraphs may be required. In these circumstances hydrographs may be synthesised on the basis of past experience in other areas and applied as first approximations to the unrecorded catchment. Such devices are called *synthetic unitgraphs*.

The best known approach is due to Snyder [47] who selected the three parameters of *hydrograph base width*, *peak discharge* and *basin lag* as being sufficient to define the unit hydrograph. These are shown in Fig. 7.24.

Snyder considered the catchment characteristics likely to affect unit hydrograph shape as being catchment area, shape of basin, topography, channel slopes, stream density and channel storage. He eliminated all these parameters except the first two by including

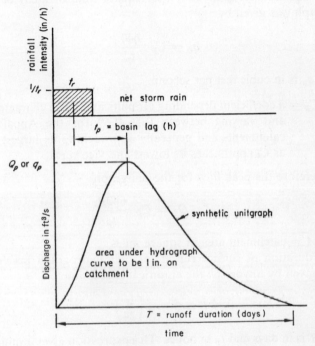

FIG. 7.24 *Synthetic unitgraph parameters*

them in a coefficient C_t. He dealt with the size and shape of catchment by measuring the length of the main stream channel and he proposed that

$$t_p = C_t(L_{ca}L)^{0\cdot3} \tag{7.3}$$

where t_p = basin lag in h, i.e. the time between mass centre of unit rain of t_r h duration and runoff peak flow.

L_{ca} = distance from gauging station to centroid of catchment area, measured along the main stream channel to the nearest point, in miles.

L = distance from station to catchment boundary measured along the main stream channel, in miles.

C_t = a coefficient depending on units and drainage basin characteristics and varying between 1·8 and 2·2 for the Appalachian Highlands catchments studied.

The equation for peak flow (per square mile of area) of the t_r unitgraph was given by

$$q_p = C_p \cdot \frac{640}{t_p} \qquad (7.4)$$

where q_p is in cubic feet per second.

C_p = a coefficient depending on units and basin characteristics and varying between 0·56–0·69 for the Appalachian catchments and generally approaching its largest value as C_t approaches its lowest and vice versa,

and therefore the peak flow for the unitgraph

$$Q_p = C_p \cdot \frac{640A}{t_p}$$

where A = catchment area in square miles.

The duration of surface runoff, or unit hydrograph baselength, T was given by Snyder by the empirical expression

$$T = 3 + 3\left(\frac{t_p}{24}\right) \qquad (7.5)$$

where T is in days and t_p in hours This expression gives a minimum baselength of 3 days for even small areas, a period much in excess of delay attributable to channel storage.

Snyder comments on this as being due to the "subsurface storm flow" which has been defined by Hursh [48] as "that portion of the storm flow which infiltrates into the surface soil but moves away from the area through the upper soil-horizons at a rate much in excess of normal groundwater seepage". This is what is referred to in recent literature as *interflow* and for most practical purposes it is regarded as surface runoff.

t_r, the unit rain period, was assumed to equal $t_p/5\cdot5$ in the study, since it was necessary to choose a single standard in all the catchment for the derivation of the formulae. This particular value was selected to make the unit of time equal to a minimum value below which further decrease would have little or no effect on basin lag or unitgraph peak discharge. If the actual length of the storm is not equal to t_r, but is t_R, the Eq. (7.4) becomes

$$q_{pR} = C_p \cdot \frac{640}{t_p + (t_R - t_r)/4} \qquad (7.6)$$

where q_{pR} = peak discharge (per square mile) of the t_R unitgraph which allows for the generally observed reduction in unitgraph peak flows with longer periods of rain. Snyder proposed subsequently [49] an expression to allow for some variation in basin lag with variation in effective rainfall duration

$$t_{pR} = t_p + (t_R - t_r)/4 \qquad (7.7)$$

where t_{pR} = basin lag for a storm of duration t_R.

Linsley subsequently presented data [50] based on a study of Californian catchments and suggested modifications of Snyder's formulae and gave values of the various coefficients, as follows:

Basin lag: a new factor t_{po} was introduced—the basin lag of an instantaneous storm—and used to derive t_{pR} which has the same meaning as before.

$$t_{po} = C_t(L_{ca}L)^{0.3} \quad \text{with} \quad C_t \text{ (average)} = 0.5$$

$$t_{pR} = t_{po} + (C_s - 0.5)t_R \qquad (7.8)$$

where C_s (average) = 0.85.

Unitgraph peak flow (per square mile): $q_{pR} = C_p \cdot 640/t_{pR}$ $\qquad (7.9)$

where C_p varies 0.35–0.50, and $Q_{pR} = q_{pR}A$

Time base of unitgraph: $T = 3 + (3t_{pR}/24)$ days

The degree of divergence of the coefficients is an indication of how important it is to attempt to obtain some actual data on the lag in an ungauged catchment, and thus enable the value of C_t to be used with reasonable assurance.

A further advance in the subject was made by Taylor and Schwarz [51] in a study based on 20 drainage basins from 20–1600 square miles in area, in north and central Atlantic states of the U.S.A.

They used the parameters L, L_{ca} as before and also introduced the slope of the main watercourse by defining S_{st} as the slope of a uniform channel having the same length as the longest watercourse and an equal time of travel. The equations derived are as follows:

Basin lag

$$t_{pR} = C'e^{m'}t_R \qquad (7.11)$$

where t_{pR} = lag in h from centroid of nett rain to hydrograph peak
t_R = time in h from beginning to end of nett rain
e = 2·7183
m' = rate of change of lag with storm duration
C' = lag of instantaneous unit hydrograph
m' and C' are derived from the following equations

$$m' = 0·212/(LL_{ca})^{0·36} \qquad (7.12)$$

$$C' = 0·6/\sqrt{(S_{st})} \qquad (7.13)$$

where L and L_{ca} have the same definitions as previously and

$$S_{st} = \left[\frac{n}{(1/S_1^{\frac{1}{2}} + 1/S_2^{\frac{1}{2}} + \cdots + 1/S_n^{\frac{1}{2}})}\right]^2 \qquad (7.14)$$

where n = Manning's coefficient of roughness for the natural watercourse

S_1, S_2 etc. = the slopes of individual sections, of equal length, into which the main watercourse may be conveniently divided.

Peak discharge (per square mile) of unitgraph

$$q_{pR} = C''e^{m''}t_R \qquad (7.15)$$

where

$$C'' = 382(LL_{ca})^{-0·36} \qquad (7.16)$$

$$m'' = 0·121S_{st}^{0·142} - 0·05 \qquad (7.17)$$

Base width of unitgraph

$$T = 5(t_{pR} + t_R/2) \text{ h} \qquad (7.18)$$

The authors give a nomogram in their paper for the solution of equations and make a number of observations about the use of their method. These include the suggestions that major tributaries should be treated separately, and that the equations should be limited to the results of moderate and major storms of uniform distribution over geographical areas similar to those from which they were derived. The equations are given here as they form a useful addition to the literature on synthesis of unitgraphs.

Unit hydrographs may also be synthesised by the methods of streamflow routing and Section 8.6 describes such a technique. It has been placed in Chapter 8 because a knowledge of routing procedure is essential to its understanding.

The coefficient n in either form of Manning's equation (British units or SI) has the same numerical value. Some typical values are listed below.

Type of channel	n
Rough-paved trapezoidal channel or canal without obstructions or sharp curves	0·021
Natural stream channel, smoothly flowing in very clean condition	0·030
Standard natural stream or river in stable condition	0·035
Rivers with shallows and meanders and noticeable aquatic growth. Streams with gravel banks and meanders	0·045–0·050
Slow flowing meandering rivers with pools and slight rapids and overgrown banks	0·060–0·100

8 Flood Routing

8.1 Introduction

Civilisation has always developed along rivers, whose presence guaranteed access to and from the sea coast, irrigation for crops, water supplies for urban communities and latterly power development and industrial water supply. The many advantages have always been counterbalanced by the dangers of floods and in the past *levees* or flood banks were built along many major rivers to prevent inundation in the flood season. In more recent times storage reservoirs have been built as the principles of dam construction became better understood and other measures like relief channels, storage basins and channel improvements are continually under construction in many parts of the world. It is important for such works that estimates can be made of how the measures proposed will affect the behaviour of flood waves in rivers so that economic solutions may be found in particular cases. *Flood routing* is the description applied to this process. It is a procedure through which the variation of discharge with time at a point on a stream channel, may be determined by consideration of similar data for a point upstream. In other words it is a process which shows how a flood wave may be reduced in magnitude and lengthened in time (*attenuated*) by the use of storage in the *reach* between the two points.

8.2 The storage equation

Since the methods of flood routing depend on a knowledge of storage in the reach, a way of evaluating this must be found. There are two ways of doing this. One is to make a detailed topographical and hydrographical survey of the river reach and the riparian land and so determine the storage capacity of the channel at different levels. The other is to use the records of past levels of flood waves at the two points and hence deduce the reach's storage capacity. It is assumed that such storage capacity will not change substantially in time and so may be used to route the passage of larger and more critical, predicted floods. As much data as possible is required for the second method, which is the one generally used, including flow

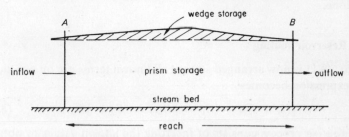

FIG. 8.1 *Storage in a river reach*

records at the beginning and end of the reach and on any tributary streams joining it, and rainfall records over any areas contributing direct runoff to it.

Storage in the reach of a river is divided into two parts, *prism* and *wedge* storage, which are illustrated in Fig. 8.1. This is simply because during floods the slope of the surface is not uniform. (See Section 6.2).

If the continuity of flow through the reach shown in Fig. 8.1 is now considered, it is clear that what enters the reach at point A must emerge at point B, or temporarily move into storage.

$$I = D + \frac{\mathrm{d}S}{\mathrm{d}t}$$

where I = inflow to the reach

D = discharge from the reach

$\dfrac{\mathrm{d}S}{\mathrm{d}t}$ = rate of change in reach storage with respect to time.

This equation is approximated to, for a time interval t, by

$$\frac{I_1 + I_2}{2} t - \frac{D_1 + D_2}{2} t = S_2 - S_1 \tag{8.1}$$

where subscripts 1 and 2 denote values at the beginning and end of the time t, respectively. The time t is called the *routing period* and it must be chosen sufficiently short, so that the assumption implicit in Eq. (8.1), i.e. that the inflow and outflow hydrographs consist of a series of straight lines, does not depart too far from actuality. In particular, if t is too long it is possible to miss the peak of the inflow curve, so the period should be kept shorter than the travel time of the flood wave crest through the reach. On the other hand, the shorter the routing period the greater the amount of computation to be done.

8.3 Reservoir routing

If Eq. (8.1) is now arranged so that all known terms are on one side, the expression becomes

$$\tfrac{1}{2}(I_1 + I_2)t + (S_1 - \tfrac{1}{2}D_1 t) = (S_2 + \tfrac{1}{2}D_2 t) \qquad (8.2)$$

The routing process consists of inserting the known values to obtain $S_2 + \tfrac{1}{2}D_2 t$ and then deducing the corresponding value of D_2 from the relationship connecting storage and discharge. This method was first developed by L. G. Puls of the U.S. Army Corps of Engineers.

The simplest case is that of a reservoir receiving inflow at one end and discharging through a spillway at the other. In such a reservoir it is assumed there is no wedge storage and that the discharge is a function of the surface elevation, provided that the spillway arrangements are either free-overflow or gated with fixed gate openings. Reservoirs with sluices may be treated as simple reservoirs also if the sluices are opened to defined openings at specified surface-water levels, so that an elevation-discharge curve may be drawn. The other required data are the elevation-storage curve of the reservoir and the inflow hydrograph.

Example 8.1. *An impounding reservoir enclosed by a dam has a surface area which varies with elevation as shown by the relationship of Fig. 8.2(a). The dam is equipped with 2 circular gated discharge ports, each of 2·7 m diameter, whose centres are at elevation 54·0, and a free overflow spillway 72·5 m long with crest level at elevation 66·0. The discharge gates are open and surface water level is at elevation 63·5, at time $t = 0$. The flood hydrograph of column 3, Table 8.2, is forecast. What will the maximum reservoir level be and when will it occur?*

1. Assume the ports have a coefficient of discharge, $C_d = 0\cdot8$, then $Q = 2(C_d A\sqrt{(2gH)})$ and at time 0, $Q = D$ (column 5 of Table 8.2) $= 2(0\cdot8 \times 5\cdot7 \times \sqrt{186\cdot5}) = 125$ m³/s. (Note $g = 9\cdot81$ m/s².) Insert this value on first line of column 5.

2. Compute the elevation–storage curve of Fig. 8.2(b). Remember that live storage starts at $52\cdot65$, the invert level of the discharge ports, and amounts to $5\cdot5 \times 10^6$ m³ by $54\cdot0$. The storage between $54\cdot0$ and $56\cdot0$ = mean area of reservoir between these levels × 2 m = $8\cdot33 \times 10^6$ m³. Successive increments, computed in this way, are plotted cumulatively with $52\cdot65$ as datum.

3. Compute the elevation–D tabulation of Table 8.1 below.

TABLE 8.1. *Elevation–discharge table*

Elevation of water surface m	Head over 54·0 m	Discharge from gated ports m³/s	Head over 66·0 = H m	$H^{\frac{3}{2}}$	Spillway discharge m³/s	Total discharge m³/s
58·0	4·0	81·0	—	—	—	81
60·0	6·0	99·5	—	—	—	100
62·0	8·0	114·8	—	—	—	115
64.0	10·0	128·0	—	—	—	128
66·0	12·0	140·7	0	0	0	141
66·1	12·1	141·2	0·1	0·032	5·1	148
66·2	12·2	141·8	0·2	0·089	14·2	156
66.3	12·3	142·6	0·3	0·164	26	169
66·4	12·4	143·0	0·4	0·252	40	183
66·5	12·5	143·7	0·5	0·353	56	200
66·7	12·7	144·8	0·7	0·58	93	238
66·9	12·9	146·0	0·9	0·85	136	282
67·0	13·0	146·4	1·0	1·0	160	306
67·5	13·5	149·3	1·5	1·84	294	443
68·0	14·0	152·0	2·0	2·83	453	605

Assume Q *spillway* $= CLH^{\frac{3}{2}}$ and use $C = 2\cdot2$ m$^{\frac{1}{2}}$/s.

4. From Fig. 8.2(b) and Table 8.1, the D–Storage curve of Fig. 8.3 may now be drawn, i.e. the central curve. The abscissa of Fig. 8.3 is graduated in "storage units". Each storage unit = routing period × 1 m³/s. Since the forecast hydrograph of column 3, Table 8.2, is given at 6 hour intervals it is convenient to make this the routing period. Then each storage unit = $6 \times 3600 \times 1 = 21\cdot6 \times 10^3$ m³ $= \frac{1}{4}$ m³/s day. The use of

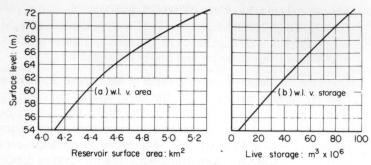

FIG. 8.2 *Reservoir characteristics*

these units is necessary to keep the dimensions of columns 4, 6 and 7 in Table 8.2 compatible.

The $S \pm \frac{1}{2}Dt$ curves are now added on either side of the storage curve. Since the abscissa is in storage units, then $t = 1$ and the curves may be plotted without calculation, e.g. at $D = 200$, $\frac{1}{2}Dt = 100$ and so two points may be set off 100 storage units on either side of the S curve, and similarly for other points.

5. The routing calculation can now be started in Table 8.2. The figures in heavy type are known.

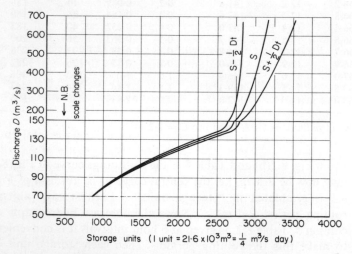

FIG. 8.3 *Reservoir routing storage curves*

TABLE 8.2 *Reservoir routing computation*

1	2	3	4	5	6	7	8
Time h	Routing period	Inflow m³/s	$\dfrac{(I_1 + I_2)t}{2}$	D m³/s	$S-\frac{1}{2}Dt$ $\frac{1}{4}$ m³/s day	$S+\frac{1}{2}Dt$ $\frac{1}{4}$ m³/s day	Surface level m
0	1	50	62	**125**	2095	2157	63·4
6	2	75	127	122	2035	2162	63·0
12	3	180	265	122	2040	2305	63·0
18	4	350	400	127	2178	2578	63·8
24	5	450	485	136	2442	2927	65·1
30	6	520	512	200	2727	3239	66·5
36	7	505	475	425	2814	3289	67.4
42	8	445	402	460	2829	3231	67·5
48	9	360	325	416	2815	3140	67·35
54	10	290	270	347	2793	3063	67·15
60	11	250	230	288	2775	3005	66·95
66	12	210	192	242	2763	2955	66·7
72	13	175	157	208	2747	2904	66·55
78	14	140	125	190	2714	2839	66·45
84	15	110	97	165	2674	2771	66·25
90	16	85	75	144	2627	2702	66·05
96	17	65	60	140	2562	2622	66·0
102	18	55	52	138	2484	2536	65·3
108	19	50	47	134	2402	2449	64·7
114	20	45	42	132	2317	2359	64·3
120	21	40	39	129	2230	2269	64.0
126		38		127			63·7

To begin with, compute column 4 by averaging successive pairs of inflow values. Now find from the $S - \frac{1}{2}Dt$ curve of Fig. 8.3, the value of this parameter corresponding to $D = 125$ m³/s. The value is 2095 and this figure is inserted in the first space of column 6.

6. The figure in column 4 is added to column 6 and the total put in column 7 (e.g. 2095 + 62 = 2157). The left side of Eq. (8.2) has now been evaluated. Find the column 7 value on the $S + \frac{1}{2}Dt$ curve and read off the corresponding value of D, entering it in column 5 (e.g. from the $S + \frac{1}{2}Dt$ curve find $D = 122$ corresponding to 2157).

7. Use this new value of D to find $S - \frac{1}{2}Dt$ again as in step 5. Note that this is directly obtainable without using the curve by sub-

tracting the value of D from the column 7 value in the line above (e.g. 2157 − 122 = 2035). The column 4 figure is then added to the new column 6 to get a new column 7 figure (e.g. 127 + 2035 = 2162).

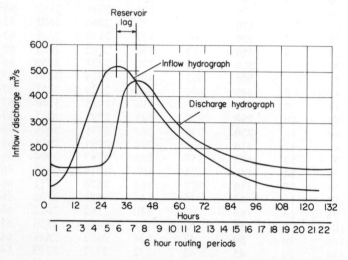

FIG. 8.4. *Inflow and discharge hydrographs for the reservoir of Example 8.1*

Complete the table and plot the outflow hydrograph (Fig. 8.4). The peak outflow should fall on the recession limb of the inflow graph.

The time difference between the peaks of the inflow and discharge hydrographs is termed *reservoir lag* and the reduction in peak flows together with the spreading out of the recession curve is referred to as *attenuation*.

8. The column 8 values of surface water level are derived from the values of discharge and levels in Table 8.1. They are most conveniently found by plotting a graph and reading off the levels corresponding to the values of column 5.

The maximum water level in the case of this example is 67·5 m occurring about hour 40.

8.4 Routing in a river channel

The solution of the storage equation in this case is more complicated than for the simple reservoir, since wedge storage is involved. Storage is no longer only a function of discharge as was the case in Example 8.1. McCarthy [52], in what has become known as the

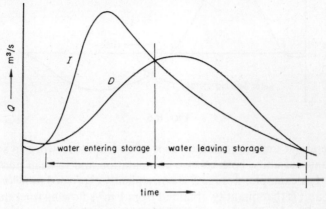

FIG. 8.5

Muskingum method, proposed that storage should be expressed as a function of both inflow and discharge in the form

$$S = K[xI + (1 - x)D] \qquad (8.3)$$

where x = dimensionless constant for a certain river reach
K = *storage constant* with dimension of time, that must be found from observed hydrographs of I and D at both stations.

The two constants may be found as follows. Let Fig. 8.5 represent the simultaneous inflow I and outflow D, of a river reach. While $I > D$, water is entering storage in the reach and when $D > I$, water is leaving it. A difference diagram can now be drawn showing this (Fig. 8.6) and subsequently, a mass curve of storage. (Fig. 8.7).

Now assume a value of x, say $x = 0.1$ and compute the value of the expression $(0.1I + 0.9D)$ for various times and plot these against corresponding S values taken from Fig. 8.7. The resulting plot, known as a *storage loop* is shown in Fig. 8.8(a); clearly there is no linear relationship. Take further values of x (say 0.2, 0.3 etc.) until a linear relationship is established, as in Fig. 8.8(c) when the particular value of x may be adopted. K is now obtained by measuring the slope of the line.

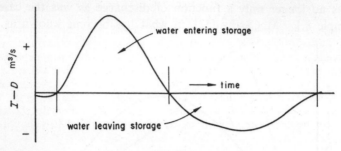

FIG. 8.6

Care about units is required. It is often helpful to work in somewhat unusual units, both to save computation and to keep numbers small. For example, storage S is conveniently expressed in m³/s day: such a unit is that quantity obtained from 1 m³/s flowing for 1 day = 86.4×10^3 m³. If S is expressed in m³/s day and the ordinate of Fig. 8.8 is in m³/s then K is in days.

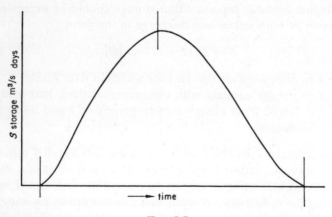

FIG. 8.7

The following excerpt from Carter and Godfrey [53] concisely sums up the choice of values for x and K:

The factor x is chosen so that the indicated storage volume is the same whether the stage is rising or falling. For spillway discharges from a reservoir, x may be shown to be zero, because the reservoir stage, and hence the storage, are uniquely defined by the outflow; hence, the rate of inflow has a negligible influence on the storage in the reservoir at any time. For uniformly progressive flow, x equals 0·50, and both the inflow and the outflow are equal in weight. In this wave no change in shape occurs and the peak discharge remains unaffected. Thus, the value of x will range from 0 to 0·50 with a value of 0·25 as average for river reaches.

The factor K has the dimension of time and is the slope of the storage-weighted discharge relation, which in most flood problems approaches a straight line. Analysis of many flood waves indicates that the time required for the centre of mass of the flood wave to pass from the upstream end of the reach to the downstream end is equal to the factor K. The time between peaks only approximates the factor K. Ordinarily, the value of K can be determined with much greater ease and certainty than that of x.

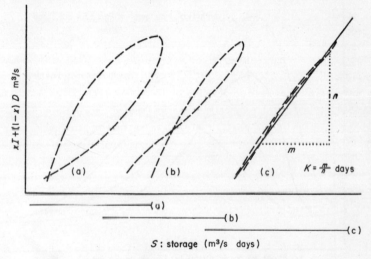

FIG. 8.8 *River routing storage loops*

Having obtained values of K and x, the outflow D from the reach may be obtained, since by combining and simplifying the two equations

$$\frac{I_1 + I_2}{2} t - \frac{D_1 + D_2}{2} t = S_2 - S_1 \qquad (8.1)$$

and

$$S_2 - S_1 = K[x(I_2 - I_1) + (1 - x)(D_2 - D_1)] \qquad (8.4)$$

(the latter being Eq. (8.3) for a discrete time interval) the equation,

$$D_2 = C_0 I_2 + C_1 I_1 + C_2 D_1 \qquad (8.5)$$

is obtained, where

$$C_0 = -\frac{Kx - 0\cdot5t}{K - Kx + 0\cdot5t}, \quad C_1 = \frac{Kx + 0\cdot5t}{K - Kx + 0\cdot5t},$$

$$C_2 = \frac{K - Kx - 0\cdot5t}{K - Kx + 0\cdot5t} \qquad (8.6)$$

where t = routing period, which should be taken as between $\frac{1}{3}$ and $\frac{1}{4}$ of the flood wave travel time through the reach (obtained from the inflow hydrograph)

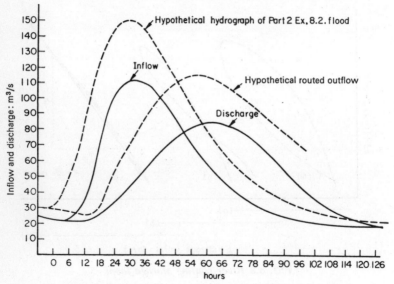

FIG. 8.9 *Inflow and discharge hydrographs for a river reach*

A worked example illustrating the application of the method is set out below.

Example 8.2. *Routing in a stream channel by the Muskingum method. Part 1*

Given the inflow and outflow hydrographs of Fig. 8.9, derive the constants x and K for the reach.

The data is set out in tabular form in Table 8.3. In columns 1 and 2, the given hydrographs are listed at a routing period interval, taken as 6 h. The storage units are taken here as ($\frac{1}{4}$ m³/s day) since the routing period is $\frac{1}{4}$ day. The columns 4, 5 and 6 are simply tabular statements of the processes illustrated in Figs. 8.6 and 8.7.

A value of x is then chosen, in the first instance 0·2, and the value inside the square brackets of Eq. (8.3) is then evaluated in columns 7, 8 and 9. Columns 6 and 9 are now plotted in Fig. 8.10 and produce the loop on the left side of the figure. (Note, incidentally, that this

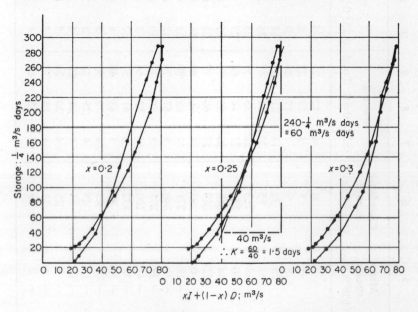

FIG. 8.10 *Storage loops for the reach of Example 8.2*

figure has ordinate and abscissa reversed compared with Fig. 8.8. This is quite unimportant and is done only for convenience of fit.)

A second value of $x = 0·25$ is now tried (columns 10–12 refer) and the resulting plot is the central one of Fig. 8.10. A third value, $x = 0·3$, is also tabulated and plotted on the right of the figure. By inspection, the central value of $x = 0·25$ approximates a straight line most nearly, so this is chosen as the x value.

K is determined by measuring the slope of the median line as shown on the figure, and is found to be 1·5 days. This confirms the

TABLE 8.3. Storage loops calculations

1	2	3	4	5	6	x = 0.2			x = 0.25			x = 0.3		
Hour	I Inflow m³/s	D Outflow m³/s	I–D m³/s	Mean storage (¼ m³/s day)	Cumulative storage (¼ m³/s day)	0.2I	0.8D Total	Total	0.25I	0.75D	Total	0.3I	0.7D	Total
0	22	22	0	0	0	4	17	21	5	16	21	7	15	22
6	23	21	2	1	1	5	17	22	6	16	22	7	15	22
12	35	21	14	8	9	7	17	24	9	16	25	10	15	25
18	71	26	45	29	38	14	21	35	18	19	37	21	18	39
24	103	34	69	57	95	20	27	47	26	25	51	31	24	55
30	111	44	67	68	163	22	35	57	28	33	61	33	31	64
36	109	55	54	60	223	22	44	66	27	41	68	33	38	68
42	100	66	34	44	267	20	53	73	25	49	74	30	46	76
48	86	75	11	22	289	17	60	77	21	56	77	26	52	78
54	71	82	−11	0	289	14	66	80	18	61	79	21	57	78
60	59	85	−26	−18	271	12	68	80	15	64	79	18	59	77
66	47	84	−37	−31	240	9	67	76	12	63	75	14	59	73
72	39	80	−41	−39	201	8	64	72	10	60	70	11	56	67
78	32	73	−41	−41	160	6	58	64	8	55	63	10	51	61
84	28	64	−36	−38	122	6	51	57	7	48	55	8	45	53
90	24	54	−30	−33	89	5	43	48	6	40	46	7	38	45
96	22	44	−22	−26	63	4	35	39	5	33	38	7	31	38
102	21	36	−15	−18	45	4	29	33	5	27	32	6	25	31
108	20	30	−10	−12	33	4	24	28	5	22	27	6	21	27
114	19	25	−6	−8	25	4	20	24	5	19	24	6	17	23
120	19	22	−3	−4	21	4	18	22	5	16	21	6	15	21
126	18	19	−1	−2	19	4	15	19	4	14	18	5	13	18

166

approximate peak to peak time of 33 h from Fig. 8.9. For this reach, therefore, use

$$x = 0 \cdot 25 \text{ and } K = 1 \cdot 5 \text{ days}$$

Part 2

Using the x and K values derived from the hydrographs, route the flood of column 2, Table 8.4, through the reach and derive the outflow hydrograph.

TABLE 8.4. *Derivation of discharge*

1	2	3	4	5	6
Hours	I m³/s	$-0 \cdot 2 I_2$ m³/s	$0 \cdot 4 I_1$ m³/s	$0 \cdot 8 D_1$ m³/s	D_2 m³/s
0	31				31·0*
6	50	− 10·0	12·4	24·8	27·2
12	86	− 17·2	20·0	21·8	24·6
18	123	− 24·6	34·4	19·7	29·5
24	145	− 29·0	49·2	23·6	43·8
30	150	− 30·0	58·0	35·0	63·0
36	144	− 28·8	60·0	50·4	81·6
42	128	− 25·6	57·6	65·3	97·3
48	113	− 22·6	51·2	77·8	106·4
54	95	− 19·0	45·2	85·2	111·4
60	79	− 15·8	38·0	89·1	111·3
66	65	− 13·0	31·6	89·0	107·6
72	55	− 11·0	26·0	86·1	101·1
78	46	− 9·2	22·0	80·9	93·7
84	40	− 8·0	18·4	74·9	85·3
90	35	− 7·0	16·0	68·3	77·3
96	31	− 6·2	14·0	61·8	69·6
102	27	− 5·4	12·4	55·7	·62·7
108	25	− 5·0	10·8	50·2	56·0
114	24	− 4·8	10·0	44·8	50·0
120	23	− 4·6	9·6	40·0	45·0
126	22	− 4·4	9·2	36·0	40·8

* assumed value

First, compute C_0, C_1 and C_2 from Eq. (8.6). Use a routing period $t = 6 \text{ h} = \frac{1}{4} \text{ day}$ as before.

$$C_0 = -\frac{(1 \cdot 5 \times 0 \cdot 25) - (0 \cdot 5 \times 0 \cdot 25)}{1 \cdot 5 - (1 \cdot 5 \times 0 \cdot 25) + (0 \cdot 5 \times 0 \cdot 25)} = -\frac{0 \cdot 25}{1 \cdot 25} = -0 \cdot 2$$

Similarly calculated, $C_1 = 0\cdot4$ and $C_2 = 0\cdot8$, check that $-0\cdot2 + 0\cdot4 + 0\cdot8 = 1\cdot0$ from Eq. (8.5)

$$D_2 = -0\cdot2I_2 + 0\cdot4I_1 + 0\cdot8D_1$$

I_1, I_2 etc. are known from the hypothetical flood hydrograph, but D_1 is unknown. Assume a value for $D_1 = I_1 = 31$ m³/s. This will be very nearly correct since the river is at a low level and will be near to a steady state. Then the first equation to be solved is

$$D_2 = -0\cdot2(50) + 0\cdot4(31) + 0\cdot8(31)$$

$$= -10\cdot0 + 12\cdot4 + 24\cdot8 = 27\cdot2$$

This value of D_2 becomes the D_1 for the next calculation and the values are tabulated as in Table 8.4.

The outflow hydrograph is plotted as a dashed line to a little way beyond the peak in Fig. 8.9.

8.5 Graphical routing methods

If Eq. (8.3) is written with $x = 0$, then

$$S = KD \qquad (8.7)$$

and since, differentiating, $\dfrac{\mathrm{d}S}{\mathrm{d}t} = K\dfrac{\mathrm{d}D}{\mathrm{d}t}$

and

$$I - D = \frac{\mathrm{d}S}{\mathrm{d}t} \text{ (from Sect. 8.2)}$$

$$\therefore \quad \frac{I - D}{K} = \frac{\mathrm{d}D}{\mathrm{d}t} \qquad (8.8)$$

This equation has been used [54] to provide a simple graphical method of routing, since $\dfrac{\mathrm{d}D}{\mathrm{d}t}$ represents the slope of the outflow hydrograph and $I - D$ and K are measurable quantities in m³/s and days. In Fig. 8.11, which is a plotted inflow hydrograph I, with discrete values I_1, I_2, I_3 etc. marked at intervals of time, the storage constant K is plotted horizontally from the position of each

I value and a line drawn from the end of the K line to the previous discharge value D. Since this line represents $\dfrac{\mathrm{d}D}{\mathrm{d}t}$, the lower part may be used to denote the actual outflow hydrograph. Naturally, the smaller the time interval the more accurate the method will be, but there is no need to have the intervals equal.

K may be varied, if its variation is known, and reference to Fig. 8.10 will suggest K may well vary and could be obtained in a relationship with outflow from such a storage loop, giving a K v. D curve as illustrated in Fig. 8.13(b).

The method can also be used in reverse, so that K at any time may be obtained from simultaneous hydrographs of I and D.

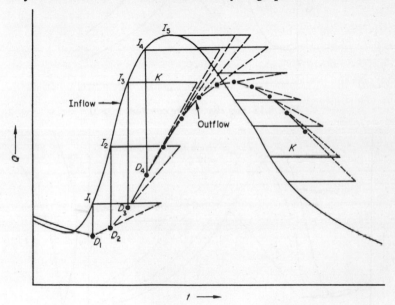

FIG. 8.11 *Graphical routing method*

The foregoing description is all qualified by the initial statement that $x = 0$, so it applies only to simple reservoir action. It may be extended, however, to include positive x values since the effect of increasing x, with K constant, is to move the outflow graph bodily to the right so that the peak value no longer falls on the recession limb of the inflow graph, and also to increase the magnitude of the outflow peak.

If a succession of historic floods is analysed, the lag that is caused by x having a positive value may be determined. The lag due to this cause, T_x, is measured from the peak of the outflow graph to the same discharge on the recession limb of the inflow graph, as illustrated in Fig. 8.12 and a plot may be made connecting T_x with corresponding I (Fig. 8.13(b)).

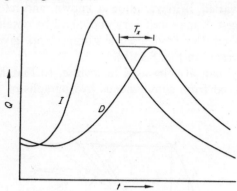

FIG. 8.12 *Lag due to the constant* $x > 0$

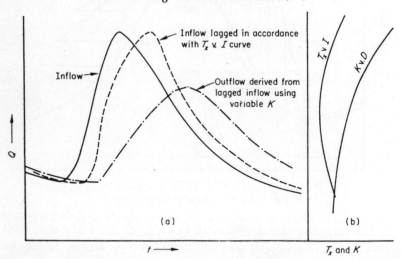

FIG. 8.13 *Graphical routing with variable lag and K*

Now the inflow graph of a reach with $x > 0$ is lagged as illustrated in Fig. 8.13, the amount of lag at each horizon being determined from the T_x v. I curve, to give the dashed inflow graph, which is then routed by the graphical method as though it were a simple reservoir inflow. It is often convenient to plot the curve of variation of K v. D alongside the $T_x–I$ curve so that both variations may be taken account of in the same plot. Fuller descriptions of this and similar routing techniques are available [31, 55, 56].

8.6 Synthetic unitgraphs from flood-routing

The principles of flood-routing may now be used to derive unit hydrographs for a catchment where almost no rainfall or runoff

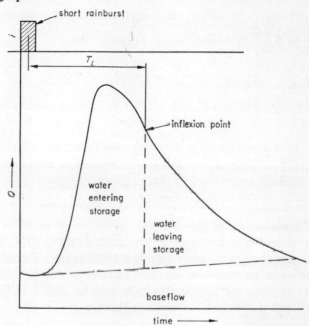

Fig. 8.14 *Hydrograph from short rain approximates IUH*

records exist. The method is not entirely synthetic since at least one observation of a runoff hydrograph must be made.

Consider a catchment as a series of sub areas, each of which, under a sudden burst of rain, contributes inflow into the system of drainage channels, which have storage. The instantaneous unit hydrograph is therefore in two parts, the first representing the inflow of the rain, and the second the gradual withdrawal of the catchment storage. The dividing line between these two parts may be conveniently taken at the point of inflexion on the recession limb as shown in Fig. 8.14.

The assumption is now made that the catchment discharge Q and the storage S are directly proportional, so that

$$S = KQ \qquad (8.9)$$

(i.e. Eq. (8.3) with $x = 0$, and Q used instead of D)

$$\text{and} \quad I - Q = \frac{\mathrm{d}S}{\mathrm{d}t} \qquad (8.10)$$

where I represents the inflow resulting from the instantaneous rain. Since $\dfrac{\mathrm{d}S}{\mathrm{d}t} = K\dfrac{\mathrm{d}Q}{\mathrm{d}t}$ by differentiating Eq. (8.9), then

$$K\frac{\mathrm{d}Q}{\mathrm{d}t} = I - Q$$

and using the condition $Q = 0$ when $t = 0$, the equation may be solved to

$$Q = I(1 - e^{-t/K}) \qquad (8.11)$$

Since the inflow ceases at the inflexion point at time T(say), then the outflow at time t in terms of discharge Q_T at T is given as

$$Q_t = Q_T e^{-(t-T)/K} \qquad (8.12)$$

The determination of the storage coefficient K must be made from an observed hydrograph on the catchment, as illustrated in Fig. 8.15, by taking two values unit time apart at the point of inflexion. The hydrograph should be of an isolated period of rain. It is not necessary that the magnitude of the rain be measured but it is necessary that it should be reasonably short, say of 1 or 2 hr only.

Then, $\quad Q_1 = Q_T$ and from Eq. (8.12) $Q_2 = Q_T e^{-(t-T)/K}$

$$\text{the shaded area } A = \int_{t=T}^{t=T+1} Q_T e^{-(t-T)/K}$$

$$= \left[-KQ_T e^{-(t-T)/K} \right]_0^1$$

$$= [K(Q_T - Q_T e^{-1/K})]$$

$$\therefore \quad A = K(Q_1 - Q_2)$$

The second observation that must be made from the observed hydrograph is the catchment lag (T_L), i.e. the maximum travel time through the catchment. This may be taken as the time from the mass centre of the causative rain (hence the requirement that it should be short—so that no large error is introduced here) to the inflexion point on the recession limb.

The storage of the catchment is now thought of as a hypothetical reservoir, situated at the point of outflow; the inflow is expressed

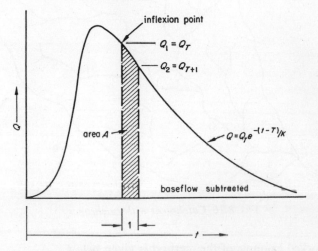

FIG. 8.15 *Determination of K*

as the time–area graph of the catchment, where each sub-area is delineated so that all rain falling on it instantaneously has the same time of travel to the outflow point, as illustrated in Fig. 8.16. The time–area graph (I) now has instantaneous unit rain applied to it and is routed through the reservoir, in the manner of Section 8.3, and the outflow (Q) derived. This outflow represents the *IUH* for the catchment and may be converted if required to the *n*-h unitgraph.

The method is basically due to Clark [57], though K is as derived by O'Kelly [58]. It is open to criticism in several respects and more advanced techniques [59, 60, 61, 62] are now available, but is has the advantage of comparative simplicity. Its derivation is not dependent on an observed hydrograph of unit intensity.

Another advantage is that instead of deriving the *IUH* (and hence the *n*-h unitgraph) design rain may be applied directly to the time–area graph, with areal variation and in any desired quantity. This produces an instantaneous design-storm hydrograph which can then be directly converted to a design-storm hydrograph of required intensity by averaging ordinates as discussed before.

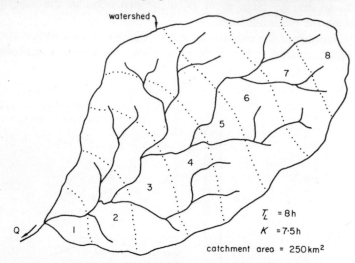

watershed

T_L = 8h

K = 7·5h

catchment area = 250 km^2

FIG. 8.16 *Catchment with isochrones*

A worked example of the method is given below.

Example 8.3. *Given the catchment area of Fig. 8.16, of area 250 km^2, and the information derived from a short rain hydrograph that T_L = 8 h and K = 7.5 h, derive the 2 h unit hydrograph*

1. Divide the catchment area into eight hourly divisions by *isochrones*, or lines of equal travel time. It will be assumed that all surface runoff falling in one of these divisions will arrive during a 1 h period at the gauging point.

2. Measure by planimeter the area of each of the hourly areas. The areas of the figure are:

Hour	1	2	3	4	5	6	7	8
Area in km^2	10	23	39	43	42	40	35	18

3. Draw the distribution graph of the runoff using the sub-areas as ordinates and 1 h intervals as abscissa. The result is the Fig. 8.17—the *time–area graph* drawn in full lines.

4. This time–area graph is now treated as the inflow I due to unit nett rain of 1 cm on the catchment of a hypothetical reservoir,

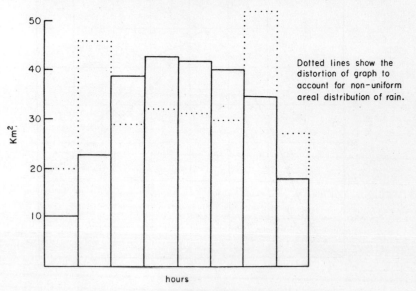

Dotted lines show the distortion of graph to account for non-uniform areal distribution of rain.

FIG. 8.17 *Sub-area distribution or time-area graph*

situated at the outlet, with storage equal to that of the catchment. Then

$$\frac{I_1 + I_2}{2} \cdot t - \frac{Q_1 + Q_2}{2} \cdot t = S_2 - S_1 \qquad \text{(from Eq. (8.1))}$$

and

$$S_1 = KQ_1 \quad \text{and} \quad S_2 = KQ_2 \quad \text{(from Eq. (8.9))}$$

From these equations

$$Q_2 = m_0 I_2 + m_1 I_1 + m_2 Q_1$$

where

$$m_0 = \frac{0\cdot5t}{K + 0\cdot5t} \qquad m_1 = \frac{0\cdot5t}{K + 0\cdot5t} \qquad m_2 = \frac{K - 0\cdot5t}{K + 0\cdot5t}$$

and since a distribution graph is being used and $I_1 = I_2$, then

$$Q_2 = m'I + m_2Q_1 \quad \text{where} \quad m' = \frac{t}{K + 0 \cdot 5t}$$

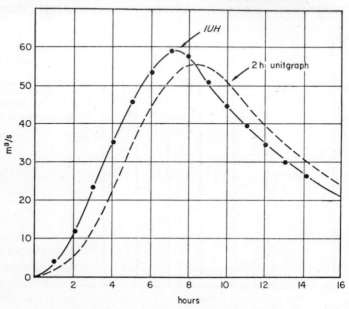

FIG. 8.18 *Derived IUH and 2 h unitgraph*

and in this case

$$m' = \frac{1}{7 \cdot 5 + 0 \cdot 5} = \frac{1}{8} = 0 \cdot 125$$

and

$$m_2 = \frac{7 \cdot 5 - 0 \cdot 5}{7 \cdot 5 + 0 \cdot 5} = \frac{7}{8} = 0 \cdot 875$$

$$\therefore \quad Q_2 = 0 \cdot 125I + 0 \cdot 875Q_1$$

5. Tabulate the data and compute Q_2 as in Table 8.5. Q_2 is the required synthetic instantaneous unit hydrograph. Compute the conversion constant for column 3.

$$1 \text{ cm rain on } 1 \text{ km}^2 \text{ in } 1 \text{ h} = \frac{10^6 \times 10^{-2}}{3600} = 2 \cdot 78 \text{ m}^3/\text{s}$$

TABLE 8.5. *IUH by routing*

1	2	3	4	5	6
Time h	Time-area diagram km²	$0.125I = 2.78$ $\times 0.125 \times$ col. 2 m³/s	$0.875 \times$ col. 5 m³/s	$Q_2 =$ col. 3 + col. 4 $= IUH$ m³/s	2 h unitgraph m³/s
0	0	0	0	0	0
1	10	3·5	0	3·5	
2	23	8·0	3·1	11·1	5·5
3	39	13·5	9·7	23·2	
4	43	14·9	20·3	35·2	23·1
5	42	14·6	30·8	45·4	
6	40	13·9	39·6	53·5	44·3
7	35	12·1	46·8	58·9	
8	18	6·2	51·4	57·6	55·5
9	0	0	50·5	50·5	
10	0	0	44·1	44·1	50·8
11	0	0	39·6	39·6	
12	0	0	34·6	34·6	39·3
13	0	0	30·2	30·2	
14	0	0	26·4	26·4	30·5
15	0	0	etc.	etc.	etc.

6. Plot the *IUH* and 2 h unitgraphs of columns 5 and 6 as Fig. 8.18.

To illustrate the ease of the method in accommodating areal variation in rainfall, suppose that a hypothetical rainfall is specified of 20 mm on sub-areas 1 and 2, 7·5 mm on 3, 4, 5 and 6, and 15 mm on areas 7 and 8, all falling in 1 h. The technique used then simply converts the time area graph by these proportions, as shown by the dotted graph lines of Fig. 8.17 before the routing operation derives the *IUH* as before and then converts it to an *n* h unitgraph by averaging each pair of ordinates at *n* h spacing. In this latter case a degree of licence is being employed using the terms *IUH* and *unitgraph* since the rainfall is not uniform and catchment-wide as required by their definition.

9 Hydrological Forecasting

9.1 Introduction

In the previous chapters the various physical processes involved in the hydrologic cycle have been enumerated and examined in detail. Methods of evaluating each process have been suggested and often explained, and techniques discussed which may be used to provide quantitative answers to many questions.

The remaining problem which must now be tackled is how to use this knowledge to predict from existing data, however meagre it may be, what will happen in future. This is a fundamental problem of all engineering design, since the engineer designs and constructs work to provide for future needs, whether he be a structural engineer designing an office block, an electrical engineer designing power systems to meet future electrical demand, or a hydraulic engineer designing reservoirs to meet future demand for water.

There is one major difference in these three cases. The structural designer is working with homogeneous materials whose behaviour is known within narrow limits. His buildings will be used by people, whose spacing, dimensions, weight and behaviour, *en masse*, can be predicted quite accurately. He has to cope with natural events only in the form of wind loads, which usually form a small proportion of the total load, and earthquakes. For both these eventualities there are codes of practice and recommendations available to him.

The electrical system designer has to extrapolate the rising demand curve of recent years, and to examine the trends of industry and personal habits, to decide how much capacity should be available in future years. While this is a complex and continuous task, it is almost completely immune from natural events other than disasters for which he cannot be expected to provide.

The hydraulic engineer on the other hand is dealing, in reservoir design, almost exclusively with natural events: in the incidence of precipitation, evaporation and so on. These events are usually *random* in nature and may have any of all the non-negative values. It is true that if the rainfall at a place is measured daily for a period of time, a knowledge about what is a probable daily rainfall will be built up, but it will not, however long it goes on, lead to any limiting possible value of daily rainfall, other than intuitively.

The hydrologist is frequently asked what the maximum possible discharge of a particular river will be. There is simply no such value. The only answer that can be given is that from the data available, and making various assumptions, it would appear that a certain value will not be exceeded on average more than once in a specific number of years. On such estimates all hydrologic design must be performed, and this chapter deals with methods whereby some of the uncertainties may be removed or narrowed in range.

9.2 Flood formulae

The particular random variable of river flood discharge has been of interest to engineers and hydrologists from the earliest days of hydrology and many formulae have been proposed to define the "maximum flood" that could occur for a particular catchment. The formulae are empirical by nature, derived from observed floods on particular catchments and usually of the form $Q = CA^n$

where Q = flood discharge in m^3/s (or ft^3/s)

A = catchment area in km^2 (or $mile^2$)

n = an index usually between 0.5 and 1.25

C = a coefficient depending on climate, catchment and units.

An early example of such a formula due to Dickens was developed in India

$$Q = 825a^{0.75}$$

with Q in ft^3/s and a in square miles; but since the formula takes no account of soil moisture, rainfall, slope, altitude etc., it is clearly of very little value in general application. This is true of all such formulae although they are frequently used to obtain a quick first estimate of the order of "maximum flood" that may be expected. For such purposes Morgan [63] proposed the formula for a catastrophic flood in Scotland and Wales of

$$Q = 3000 M^{0\cdot5}$$

where Q is in ft³/s and M is catchment area in square miles, and added the sophistication of a recurrence period T (in years) by quoting

design flood = catastrophic flood $\times (T/500)^{\frac{1}{4}}$

for cases where the adoption of the catastrophic flood was not justified by danger to human life or the safety of a dam. A similar formula of the same type, due to Fuller, has been widely used in the U.S.A.

$$Q_{av} = CA^{0\cdot8}$$

where A is catchment area in square miles

C is a coefficient often taken as 75

Q_{av} is average value of annual flood discharge in ft³/s

The value of Q_{av} is then substituted in the formula

$$Q_m = Q_{av}(1 + 0\cdot8 \log T)$$

where T is a return period in years and Q_m is the 'most probable' annual maximum flood. Such calculations, while simple to make, are of limited value, in that they are 'envelope' expressions derived to cover all recorded occurrences with indeterminate safety margins and as such they take no account of the physical processes involved in runoff, and are frequently very conservative.

9.3 Frequency analysis

9.3.1 Series of events.
The next approach is to use the methods of statistics to extend the available data and hence predict the likely frequency of occurrence of natural events. Given adequate records, statistical methods will show that floods of certain magnitudes may, on average, be expected annually, every ten years, every 100 years and so on. It is important to realise that these extensions are only as valid as the data used. It may be queried whether *any* method of extrapolation to 100 years is worth a great deal when it is based on (say) 30 years of record. Still more does this apply to the '1000 yr flood' and similar estimates.

Another point for emphasis is the non-cyclical nature of random events.* The 100 yr flood, (i.e. the flood which will occur on *average*, once in 100 years) may occur next year, or not for 200 years or may be exceeded several times in the next 100 years. The accuracy of estimation of the value of the (say) 100 yr flood depends on how long the record is and, for flood flows, one is fortunate to have records longer than 30 years. Notwithstanding these warnings, frequency analysis can be of great value in the interpretation and assessment of events such as flood flows and the risks of their occurrence in specific time periods.

It is particularly important to define what is meant by an event. For example if a river has been gauged every day for 10 years, there

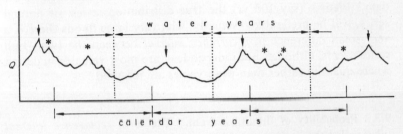

FIG. 9.1 *Annual and partial duration series events*

will be about 3650 observations. These are not independent random events since the flow on any one day is dependent to some extent on that of the day before, and so the observations do not comprise an *independent series*. The array of these observations is termed a *full series*.

Suppose from the 10 year record we extract in each year the maximum event. These would constitute an independent series since it is highly unlikely that the maximum flow of one year is affected by that of a previous year. Even so, care is necessary, as may be seen from Fig. 9.1, where *water years*, measured between seasons of minimum flow, are marked, as well as calendar years. One calendar year might contain two water-year peaks, so it is necessary to specify that water years should be used in defining events. A selection like

* See also Section 9.7.

this is called an *annual series*. Such a series is open to the objection that some of the peaks are smaller events than secondary peaks (marked with an asterisk in Fig. 9.1) of other years. The objection may be overcome by listing a *partial duration series*, in which strict time segregation is no longer a condition and all peaks above some arbitrary value (say the lowest annual peak) are included provided that, in the judgment of the compiler, they are independent events, uninfluenced by preceding peak flows. Partial duration series events therefore allow the objection of subjective judgment and are not, strictly speaking, independent and random.

Which series is used depends on the purpose of the analysis. For information about fairly frequent events, e.g. the size of a flood that might be expected during the construction period of a large dam (4 years say), then a partial series may be best, while for the design flood for the dam's spillway which should not be exceeded in the dam's lifetime (say 100 yr) the true distribution series, or annual series, will be preferable. Actually, with very large floods there is a very small difference in *recurrence interval* between the two. Full series events, although not independent, are most valuable in design where quantity rather than peak values are required.

9.3.2 Probability of the N-yr event.

The term *recurrence interval* (also called the *return period*), denoted by T_r, is the time which, on average, elapses between two events which equal or exceed a particular level. Putting it another way, the N-yr event, the event which is expected to be equalled or exceeded, on average, every N years, has a recurrence interval, T_r of N years.

As mentioned previously there is no implication that the N-yr event occurs cyclically. It does, however, have a probability of occurrence in any particular period under consideration.

Let the probability $P(X \leqslant x)$ represent the probability that x will not be equalled or exceeded in a certain period of time.

Then $P(X \leqslant x)_n$ will represent the probability that x will not be equalled or exceeded in n such periods.

For an independent series and from the multiple probability rule

$$P(X \leqslant x)_n = [P(X \leqslant x)]^n$$
$$= [1 - P(X \geqslant x)]^n$$
$$\therefore \quad P(X \geqslant x)_n = 1 - [1 - P(X \geqslant x)]^n$$

Now
$$T_r = \frac{1}{P(X \geqslant x)}$$

$$\therefore \quad P(X \geqslant x)_n = 1 - \left[1 - \frac{1}{T_r}\right]^n$$

So, for example, the probability of $X \geqslant x$, where x is the value of a flood with a return period of 20 years, occurring in a particular 3 year period is

$$
\begin{aligned}
P(X \geqslant 20 \text{ yr flood})_3 &= 1 - [1 - \tfrac{1}{20}]^3 \\
&= 1 - [0.95]^3 \\
&= 1 - 0.857 \\
&= 0 \cdot 143 \text{ or } 14 \cdot 3\%
\end{aligned}
$$

Table 9.1 shows the probability of the N-yr flood occurring in a particular period.

TABLE 9.1 *Percentage probability of the N-year flood occurring in a particular period*

Number of years in period	N = Average return period T_r: years							
	5	10	20	50	100	200	500	1000
1	20	10	5	2	1	0·5	0·2	0·1
2	33	19	10	4	2	1	0·4	0·2
3	45	27	14	6	3	1·5	0·6	0·3
5	63	41	22	10	5	2·5	1	0·5
10	87	65	40	18	9	5	2	1
20	98	88	64	33	17	10	4	2
30	99·8	96	78	45	24	14	6	3
60	—	99·8	95	70	43	26	11	6
100	—	—	99·4	87	60	39	18	9
200	—	—	—	98·2	74	63	33	18
500	—	—	—	—	99·6	92	63	39
1000	—	—	—	—	—	99·3	96	63

Where no figure is inserted the % probability $> 99 \cdot 9$.

For example, it may be seen from the table there is a 1% chance of the 200 yr flood occurring in the next 2 years and an 8% chance that it will not occur for the next 500 years.

If the probability $P(X \leqslant x)_n$ is defined by a policy ruling, the value of n, the design period, may be found from

$$P(X \geqslant x)_n = 1 - \left(1 - \frac{1}{T_r}\right)^n$$

$$1 - P(X \geqslant x)_n = \left(1 - \frac{1}{T_r}\right)^n = \left(\frac{T_r - 1}{T_r}\right)^n$$

$$\log\left(1 - P(X \geqslant x)_n\right) = n \log\left(\frac{T_r - 1}{T_r}\right)$$

$$\therefore \quad n = \frac{\log\left(1 - P(X \geqslant x)_n\right)}{\log\left(\dfrac{(T_r - 1)}{T_r}\right)}$$

Example 9.1.   *How long may a cofferdam remain in a river, with an even chance of not being overtopped, if it is designed to be secure against a 10 year flood?*

Here, the policy ruling is that there should be an even chance, so $P(X \geqslant x)_n = 0\cdot50$ and $T_r = 10$ then

$$n = \frac{\log(1 - 0\cdot5)}{\log\frac{9}{10}} = \frac{\log 0\cdot5}{\log 0\cdot9} = \frac{\bar{1}\cdot699}{\bar{1}\cdot954} = \frac{0\cdot301}{0\cdot046} = 6\cdot5 \text{ yr}$$

9.3.3 Probability plotting.   Having listed a series of events they may each then be accorded a ranking, m, starting with $m = 1$ for the highest value, $m = 2$ for the next highest and so on in descending order.   The recurrence interval T_r (in years) of each event is now computed from

$$T_r = \frac{n + 1}{m} \tag{9.1}$$

where m = event ranking, and n = no. of events.

(Note: when plotting partial duration series, it is customary to limit the number of events to the n highest, where n = number of years of observation.)

Various other formulae are used in place of Eq. (9.1) which is most commonly used: for example, the California formula [64]

$$T_r = \frac{n}{m}$$

and Hazen's formula [65]

$$T_r = \frac{2n}{2m - 1}$$

Table 9.2 shows a listing of the annual maximum daily mean flows of the River Thames at Teddington Weir, for the years 1882–1967. This is a true annual series of random events with return periods computed.

TABLE 9.2. *Maximum mean daily flows for water years 1882–1967, for R. Thames at Teddington*

Water-year ending 30 Sept	Q_m m³/s	Rank m	Return period T_r yr	Percent probability P	Water-year ending 30 Sept.	Q_m m³/s	Rank m	Return period T_r yr	Percent probability P
1882	—	—	—	—	1925	522	7	12·3	8·1
1883	292	46	1·87	53·5	1926	370	25	3·44	29·1
1884	231	65	1·32	75·5	1927	375	23	3·74	26·8
1885	230	67	1·28	78·0	1928	526	6	14·3	7·0
1886	244	59	1·46	68·6	1929	235	62	1·39	72·0
1887	284	48	1·79	55·8	1930	552	4	21·5	4·6
1888	208	73	1·18	84·9	1931	228	69	1·25	80·3
1889	237	61	1·41	71·0	1932	274	49	1·75	57·0
1890	205	74	1·16	86·0	1933	478	9	9·55	10·5
1891	171	81	1·06	93·2	1934	95	85	1·01	99·0
1892	339	32	2·69	37·2	1935	227	71	1·21	82·5
1893	300	42	2·04	48·9	1936	478	10	8·6	11·6
1894	173	79	1·09	92·0	1937	438	14	6·15	16·3
1895	789	1	86	1·16	1938	247	58	1·48	67·5
1896	202	76	1·13	88·4	1939	369	26	3·30	30·2
1897	351	29	2·96	33·8	1940	410	15	5·74	17·4
1898	171	80	1·07	93·0	1941	384	19	4·52	22·1
1899	262	51	1·69	59·3	1942	298	44	1·95	51·2
1900	533	5	17·2	5·8	1943	457	11	7·8	12·8
1901	200	77	1·12	89·5	1944	115	83	1·04	96·5
1902	162	82	1·05	95·5	1945	261	52	1·65	60·5
1903	386	17	5·06	19·8	1946	257	53	1·62	61·6
1904	516	8	10·8	9·3	1947	714	2	43	2·3
1905	229	68	1·26	79·0	1948	227	70	1·23	81·5
1906	249	57	1·51	66·3	1949	299	43	2·00	50·0
1907	220	72	1·19	83·7	1950	324	35	2·46	40·7
1908	376	21	4·1	24·4	1951	385	18	4·78	20·9
1909	204	75	1·15	87·1	1952	377	20	4·3	23·2
1910	231	66	1·30	76·7	1953	263	50	1·72	58·1
1911	395	16	5·38	18·6	1954	231	64	1·34	74·5
1912	367	28	3·07	32·6	1955	453	13	6·61	15·1
1913	255	55	1·57	64·0	1956	316	38	2·26	44·2
1914	256	54	1·59	62·8	1957	314	39	2·20	45·4
1915	585	3	28·6	3·5	1958	317	37	2·32	43·0
1916	373	24	3·58	27·9	1959	375	22	3·91	25·6
1917	327	34	2·53	39·6	1960	308	41	2·10	47·7
1918	351	30	2·86	34·9	1961	456	12	7·16	14·0
1919	334	33	2·60	38·4	1962	344	31	2·78	36·0
1920	251	56	1·54	65·1	1963	286	47	1·83	54·6
1921	240	60	1·43	69·8	1964	369	27	3·18	31·4
1922	198	78	1·10	90·6	1965	113	84	1·02	97·6
1923	231	63	1·37	73·2	1966	324	36	2·39	41·9
1924	298	45	1·91	52·4	1967	313	40	2·15	46·5

$n = 85$ $Q_{av} = 319·5$ m³/s   standard deviation $\sigma = 124·56$ $T_r = (n + 1)/m$ $P\% = 100/T_r$

Having obtained the T_r values, the question now arises, can the listed values of discharge, Q and T_r be used to extrapolate the data and infer the return periods of extreme floods? This may be attempted in various ways, which are discussed below.

(a) Q v. T_r may be plotted on simple plane co-ordinates. An example of this is illustrated in Fig. 9.2 where the River

Thames data of Table 9.2 has been plotted. The extrapolation of the curve to high Q and T_r values depends almost entirely on the few highest existing points.

(b) Q linear and T_r logarithmically. The same data has now been plotted in Fig. 9.3 and a straight line fitted to it. Since the judgment required in this case is to fitting a line through all the points rather than extrapolating from a few, it is simpler

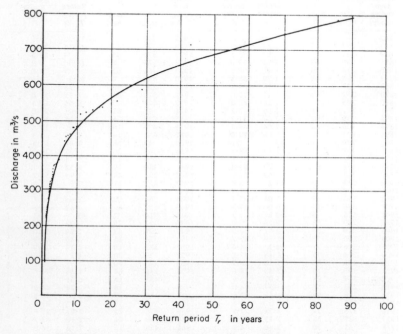

FIG. 9.2 *Annual maximum daily flow of River Thames at Teddington, 1882–1967*

to do than the first case. However, unless the return period follows a logarithmic law, it does not necessarily provide more accurate extrapolation.

(c) Another approach is to assume that the events have a *normal distribution*, so that on *normal probability* paper (due to Hazen [66] they will plot along a straight line. This plot is shown in Fig. 9.4. Clearly the points do not lie along a straight line, so a shallow curve has been fitted to them. It should be noted that the T_r scale has now become probability P. P is

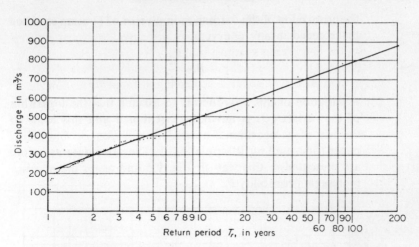

FIG. 9.3 *Annual maximum daily flow of River Thames at Teddington, 1882–1967 (semi log)*

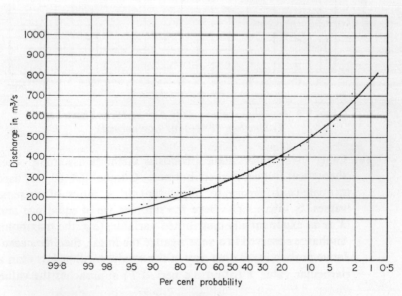

FIG. 9.4 *Annual maximum daily flow of River Thames at Teddington, 1882–1967 (normal probability)*

187

the probability of the corresponding value of Q being equalled or exceeded in any one year.

$$P = \frac{1}{T_r}$$

(d) A variation of this approach is to assume that the logarithm of the variate (Q) is normally distributed and this leads to the use of a logarithmic-normal distribution or *log-normal* paper (first used by Whipple [67].) The same data as before are presented in this way in Fig. 9.5 and again the abscissa represents probability P.

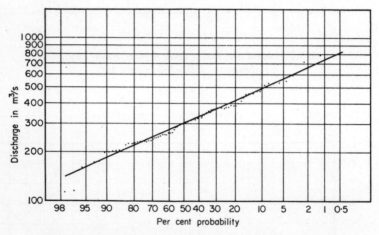

FIG. 9.5 *Annual maximum daily flow of River Thames at Teddington, 1882–1967 (log normal)*

(e) Other investigators have proposed methods assuming other theoretical frequency distributions. Gumbel [68, 69, 71] used *extreme-value theory* to show that in a series of extreme values $X_1, X_2 \ldots X_n$ where the samples are of equal size and X is an exponentially distributed variable, (e.g. the maximum discharge observed in a year's gauge readings), then the cumulative probability P' that any of the n values will be less than a particular value X (of return period T) approaches the value

$$P' = e^{-e^{-y}}$$

where e = natural logarithm base

and
$$y = -\ln\left[-\ln\left(1 - \frac{1}{T}\right)\right]$$

i.e. P' is the probability of non-occurrence of an event X in T years.

or
$$T = \frac{1}{1 - P'}$$

(Note: This argument refers to Gumbel's method. The reader should not confuse this with the normal usage of $T_r = \dfrac{1}{P}$ where P = probability of occurrence.)

The event X, of return period T years, is now defined as Q_T,

then
$$Q_T = Q_{av} + \sigma(0{\cdot}78y - 0{\cdot}45) \qquad (9.2)$$

where Q_{av} = average of all values of 'annual flood' Q_m
and σ = standard deviation of the series

$$= \sqrt{\left[\frac{n}{n-1}\left(\frac{\Sigma Q_m^2}{n} - Q_{av}^2\right)\right]}$$

where n = number of years of record = number of Q_m values

and ΣQ_m^2 = sum of the squares of n values of Q_m

Table 9.3 gives values of y as a function of T.

TABLE 9.3

T	y	T	y	T	y
1·01	− 1·53	5·00	1·50	50	3·90
1·58	0·00	10·00	2·25	100	4·60
2·00	0·37	20·00	2·97	200	5·30

Powell [70] suggested that if plotting paper is prepared in which the horizontal lines are spaced linearly and the vertical lines' spacing is made proportional to y, then from Eq. (9.2) Q_T and T will plot as straight lines. This is the basis of Gumbel-Powell probability paper, used to plot the River Thames data in Fig. 9.6. The return period T has been computed, as before, as $T = \dfrac{n + 1}{m}$. The straight line on this figure has been drawn between the two points Q_{av} and Q_{200}. Q_{av}, from Eq. (9.2) occurs when $0\cdot78y = 0\cdot45$ or $y = 0\cdot577$, which corresponds to $T = 2\cdot33$ yr. Eq. (9.2) holds for large values of n, say $n > 50$, when Q_{av} at $2\cdot33$ yr is included on the line through the points. The other point Q_{200}, represents the '200-yr flood' and is found by inserting the appropriate values in Eq. (9.2).

$$Q_{200} = Q_{av} + 124\cdot56(0\cdot78 \times 5\cdot30 - 0\cdot45) = 778 \text{ m}^3/\text{s}$$

The correspondence between the plotted data and Gumbel's theoretical line is demonstrated. Gumbel paper should not be used for

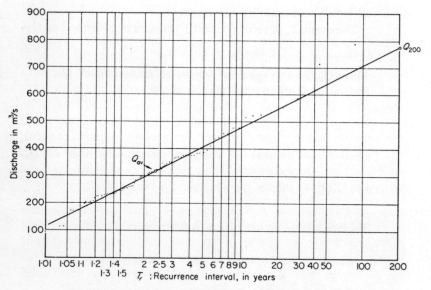

FIG. 9.6 *Annual maximum daily flow of River Thames at Teddington, 1882–1967 (Gumbel distribution)*

partial series which usually plot better on semi-log paper, as used in Fig. 9.3.

From the plots presented in Figs. 9.2–9.6, it may seem there is little to choose between the particular plotting papers available. This is very often the case and investigators should use whichever distribution makes their particular job of fitting and extrapolation simplest and the line apparently of best fit.

The foregoing is a necessarily brief résumé of the methods of plotting flood events, in current use. For the underlying theories the reader should refer to original papers and more comprehensive treatments available [72, 73].

9.4 Synthetic data-generation

One of the perennial problems of the hydrologist is insufficient data, whether it be rainfall or, more usually, discharge observations. If he wishes to predict flood flows of relatively large magnitude and hence large return periods, he may find that he has perhaps a decade or two of daily observations which represent discharges. Using an annual series such a record might yield 20 or 30 points from which it is clearly dubious to predict rare events of the order of the 100 year flood. By using only one or two of these measurements from each year's record, an enormous amount of information about the discharge characteristics of catchment is being left untouched. Is there not contained within this mass of routine observation a guide, not only to the catchment's response to rain, but to the incidence of the rain itself? By studying the fundamental nature of the variations in discharge, even at low flows, might it be possible to reproduce them, aoting in their random and apparently uncoordinated ways and so automatically reproduce series exhibiting the variations that the natural data show?

The advent of powerful computers has provided answers to these questions by analysing and reproducing synthetic data in large quantities. While the quality of the synthetically generated data is mainly dependent on the original natural data, the method makes use of all the information available rather than the very small proportion of it in the form of extreme values.

In essence, the method of artificially generating time-series relies on using historical records as a sample of a total population, while conventional methods consider the records to be the total population. It follows that designs will be based on estimates of what might have happened instead of on what has happened.

Any time series of observed values may contain a *trend* component, a *periodic* component and a *stochastic* component. The first two components are *deterministic* in nature, i.e. they are not independent of the time at which a series starts nor of the length of the series, while the stochastic component is *stationary*, i.e. the statistics of the sample do not differ from the statistics of the population (except as a result of sampling variability) and are time-independent.

If the trend and periodic components are removed from the series, a stationary stochastic component is left. This component will contain a random element and may or may not contain a correlation element. Series correlation describes how each term in a series is affected by what has gone before, e.g. a wet summer may lead to higher autumn flows than average. Accordingly, the random element and correlation structure of the stochastic component must be isolated and quantified.

By now the time series has been taken apart and its various parts examined. Each of the parts is now reproduced by mathematical simulation using randomly occurring numbers, Markov series, serial correlation coefficients etc., including the re-introduction of the periodicity and trend components. The 'model' thus created can now be used to generate synthetic data in whatever quantities are desired, and the series produced used to estimate particular N-year events as though the data had been observed.

9.5 Rainfall data and the unit hydrograph

In using unit hydrographs for hydrologic forecasting it is only necessary to decide what effective rain to apply, since the unitgraph can be adjusted to suit the chosen storm length (Sect. 7.5). This, however, is no simple task and is one of the least well documented subjects in hydrological literature. The reason for this is mainly that the operation of applying rain to unit graphs is carried out for many different reasons. The design and construction of different projects may involve the estimation of many different floods. For example, the dam designer wants to make his spillways sufficiently large so that the safety of the dam is not threatened in its lifetime. The dam builder wants to know what risk he runs of various floods occurring in the 3 or 4 years he may be building in a river channel, so that his cofferdams and tunnels may be economically sized. The flood protection engineer wishes to be sure his levees will not be overtopped more often than at some design frequency which he regards as economic, or perhaps not at all if the safety of human life is concerned. The estuarial or river port manager wants to know the depth of water he can rely on in navigable channels and so on.

In any or all of these cases it may be desirable to choose a design rain of particular frequency and magnitude and apply it to a catchment unitgraph as a means of flow prediction. This is particularly so because in many parts of the world there are reasonably long-term rainfall records though very sparse or short-term runoff records.

The rainfall records must therefore be examined and frequency analyses made of the incidence of particular 24 h rainfall depths. Rainfall observations are comparatively rarely made more often than this, although clearly very intense storms of rain frequently last for shorter periods. Much depends on catchment size. If it is very large, then 24 h frequencies may be adequate for choosing design rain. Most often, however, shorter periods of rain at greater intensity will represent the greatest flood danger and efforts must be made to find shorter interval data, or to start measuring it by autographic recorder as soon as project investigations commence. Extrapolations may be made by recording the frequencies with which certain rainfall depths occur in 96 h, 72 h, 48 h, and 24 h periods. Much data is already available, from U.S. sources [74, 75, 76, 77, 78] which though directed obviously to U.S. catchments, is applicable with care elsewhere. In Britain, reference should be made to the Flood Study Report expected to be published in 1974 [79]. The Meteorological Office contribution to this report allows estimates of expected rainfall amounts to be made for any place in the United Kingdom with a given return period for durations from 15 seconds to 30 days. It is also possible to obtain reduction factors to be applied to point rainfall to calculate areal rainfall for the same return period, for a rainfall duration from 2 minutes to 8 days and for areas between 1 and 10,000 km^2.

Reference may also be made to Bleasdale [80], to British Rainfall 1939 *et seq.* and others [81].

The process of using the unit-hydrograph therefore involves the following steps:

(a) An examination of all relevant data (including frequency analyses), in particular, any autographic rain recordings, and a subsequent selection of a total precipitation that could, in the judgment of the individual concerned, occur in a reasonably short time period, over the catchment concerned. (Note: where the catchment is large and it seems unlikely that uniform precipitation could or would occur, it should be split into a series of sub-catchments and each of these should be assigned a precipitation.)

(b) From the total precipitation, thus chosen, all losses must be subtracted. These will include interception, infiltration (having regard to the soil-moisture deficiency) and evapotranspiration. (See Section 4.4, and Chapter 3.)

(c) The resultant nett, or effective rainfall is then applied to a unit-graph adjusted to the correct time-base, and the resulting storm hydrograph ordinates obtained. If more than one hydrograph is concerned the temporal spacing of the rain must be decided, usually, but not necessarily, to obtain the maximum possible combination of ordinates.

(d) Base-flow, assessed separately, is now added to give the total storm runoff.

The choice of the s.m.d. on which a design storm may fall is particularly significant. It may not always be reasonable to presuppose a saturated catchment, and a hypothetical infiltration capacity curve, a Φ-index or coaxial correlation graph (e.g. Fig. 4.7) may be used with this in mind.

9.6 Hydro-meteorology

Over the last 25 years there has been an increasing use of the science of meteorology, in combination with hydrology, by engineers concerned with water resources and flood control. The development of this trend is due to the need to use large values of rainfall to apply to the unit hydrograph, which is a powerful tool in the estimation of extreme values of flood flows used in the design of water control structures. Immediately, the problem is met of how large is 'large'? The need to define the upper limits of possible precipitation more precisely than can be done by statistical methods (based on what are comparatively short records) led to the investigation of just how much water is stored in a storm.

Hydro-meteorology, like many other hydrological techniques, has been developed mainly in the U.S.A. and it seems reasonable, therefore, to discuss it in the light of procedures used there. The U.S. Corps of Engineers prepare flood estimates of three classes under the headings:

(a) *Statistical analyses of stream flow records.* These are made generally on a regional basis along the lines discussed in Sect. 9.2. They are used mainly in assessing mean annual benefit over the long-term, as a result of implementing particular projects.

(b) *Standard project flood (SPF) estimates.* These estimates are of the floods likely to occur from the most severe combination of meteorological and hydrological conditions which are reasonably characteristic of the geographical area being considered, but excluding extremely rare combinations. The concept is a good one but inevitably subjective. The difference between "extremely rare" and "rare" and decisions about what is "reasonable" obviously call for judgment and have caused discussion (without resolution) before [82].

(c) *Maximum-probable-flood (MPF) estimates.* These estimates differ only from the SPF in that they are reckoned to include the extremely rare event, or what has been termed in much British literature "the catastrophic flood", and are usually confined to spillway design for high dams.

The SPF and MPF estimates are examples of the continuing modern tendency to rely on the science of meteorology for the prediction of very large floods. This meteorological approach is concerned with those physical conditions of the atmosphere in particular climates, which represent the upper limits of rainfall production. The variation of these conditions in magnitude, season, temporal distribution and frequency all require consideration.

Probable maximum precipitation (PMP). The hydrometeorologist starts from the premise that if the greatest amounts of rain that can fall over a given catchment in a given time can be estimated, then the application of this rainfall with appropriate reduction for losses to the unit hydrographs of the catchment will provide an estimate of the probable maximum flood. This is likely to be a better estimate of the MPF than the extrapolation of flow records, to 20 or 30 times their period of observation, by fitting statistical distributions.

The essential requirement for precipitation is a supply of moist air. The air's water content is measured by its temperature and dew-point, which are standard meteorological observations and are therefore often available in catchments where few flow records exist. The amount of water in the atmosphere above a catchment can be established by the assumption of a saturated adiabatic lapse rate (see Sect. 2.3) or by actual weather balloon observations. The amount is usually between 10 mm and 80 mm in depth. This water is precipitated by cooling, which is accomplished almost always by vertical movement, the air mass expanding adiabatically, thus precipitating moisture which in turn liberates latent heat, which accelerates the vertical movement. The process is self-stimulated and in extreme cases is responsible for the "cloud burst" of the severe thunderstorm. The vertical lifting may also be due to orographic or frontal effects as discussed in Section 2.6.

The inflow of moist air at the base of an ascending storm column may be measured by wind-speed observations around storm peripheries. These observations yield data about the amount of water which may be carried into a storm-column and hence precipitated. The data may also be inferred from synoptic weather charts if these are based on a sufficiently dense network of observer stations.

From these observations of air-moisture, (through dew-point and temperature) and air-inflow (through wind speed and barometric pressure) a *storm moisture inflow index* may be found. This method is described in the literature for various areas of the world [78, 83, 84, 85]. Maxima for separate seasons or months may be subsequently derived by taking maximum recorded seasonal or monthly values of air-moisture and air-inflow and assuming they occur simultaneously. These may be compared with historic storms which have occurred in the catchment and meteorologically-similar neighbouring regions. In this way an array of data about the historic storms may be built up and compared with the hypothetical maxima.

The PMP may be derived from the hypothetical maxima by taking the peak of the envelope curve covering them. It is argued sometimes that the elements used to define the maxima should themselves be subject to frequency analysis before the adoption of a PMP. This, however, presupposes that the physical processes of rain producing storms are random events which can occur in vastly different magnitudes. This may not be true to the same extent as it is of precipitation recorded at a rain gauge or the flood in a river. Also, in taking the known maximum values of the determining factors and combining them in space over the catchment, the concept of return period is invalidated. The approach is deterministic rather than statistical.

The subject is under active development in many countries and more precise techniques will no doubt be developed presently.

General application of data. Many large storms have been analysed, particularly in the United States, and maximum values of rainfall depth for various durations and areas have been published [75, 76, 86]. Such data are usually presented as sets of curves, each representing a rainfall depth plotted on rectangular coordinates of storm duration in hours and storm area in square miles. These are a useful guide to limiting values but must be used with judgment for particular catchments, since topography and elevation as well as climate may modify the results appreciably for other regions.

In the light of this section it is interesting to look again at the world maximum point rainfalls quoted in Section 6.5. Paulhus [87] suggests that if rainfall is plotted against duration, both scales logarithmic, the world's greatest observed point rainfalls lie on or just under a straight line whose equation is

$$R = 16 \cdot 6 D^{0 \cdot 475}$$

Where R is rainfall in inches and D is duration in hours. Bleasdale [88] has suggested that the simpler relationship

$$R = 15 \cdot 5 D^{\frac{1}{2}}$$

is a reasonably good fit and is easier to calculate, and points out that British maxima also lie very close to a straight line on the same plotting, with values close to one quarter of the world's values.

It is probable, however, that the simple relationship quoted represents a smoothing of several overlapping complex relationships and naturally it would be unwise to use the straight line values in areas or on catchments whose records show precipitation falling well below those where the maxima have occurred.

9.7 The cyclical nature of hydrological phenomena

In all the foregoing sections of this book it has been tacitly assumed that the processes described and studied are based on non-changing physical conditions. For example in the analysis of frequencies it is assumed that events which occurred in the last 50–100 years can be used to predict the probability of similar events occurring in future. From time to time this assumption is challenged but has rarely been disproved.

This is at least partly due to misconceptions about what periodicity implies, when associated with hydrological events. The implication of a periodicity in hydrological phenomena is that the likelihood of certain values of random events appears greater at certain times than at others. In other words there is a cyclical change in probabilities, rather than events. This does not rule out the possibility of maximum events (for example) occurring at times when their probabilities are least.

Brooks and Carruthers [89] describe the periodicity of annual rainfall in England which seems to have a period of 51·7 years. They go on to show that the probability of a wet year was almost twice that of a dry year near the maximum of the 51·7 year cycle, whilst near the minimum it was less than half. This was true despite both very wet and dry years occurring at corresponding times of mini-mum probability.

It seems futile to employ statistical methods to drive probabilities of occurrence of certain events without recognising that certain processes which are part-causative agents for these events, may be subject to cyclic probability themselves, thereby altering the derived probability to some degree. Although the opportunity for using such information may be rare, its presence should be investigated for any short-term probability analysis where the time-span is of the same order as the derivative data.

Cochrane, in an analysis of the hydrology of Lake Nyasa and its catchment [90] has demonstrated a correlation between the rate of change of sunspots and the "free water" on Lake Nyasa. (The term "free water" refers to the residual runoff from the catchment and storage on the lake, after losses have been deducted from rainfall.) In another paper [91] he quotes many references to support the case for cyclic behaviour in hydrologic phenomena, concerning work in many parts of the world. A dispassionate examination of the evi-dence leads the present writer to the conclusion that periodicity in hydrological phenomena exists, however imperfectly as yet, we understand it.

Problems

Some of the following problems have appeared in the University of Salford's B.Sc. (Civil Eng.) examination papers, and are published by permission of the University.

Answers follow on p. 226.

Chapter 2—Problems

2.1 An air mass is at a temperature of 28°C with relative humidity of 70%. Determine

    (a) saturation vapour pressure
    (b) saturation deficit
    (c) actual vapour pressure in mbar and mm Hg
    (d) dewpoint
    (e) wet bulb temperature

2.2 Discuss the relationships between depth, duration and area of rainfall for particular storms.

2.3 The following are annual rainfall figures for four stations in Derbyshire. The average values for Cubley and Biggin School have not been established.

	Average (in.)	1959	1960
Wirksworth	35·5	26·8	48·6
Cubley		19·5	42·4
Rodsley	31·3	21·6	42·1
Biggin School		33·1	54·2

    (a) Assume departures from normal are the same for all stations. Forecast the Rodsley "annual average" from that at Wirksworth and the two years of record. Compare the result with the established value.

(b) Forecast annual averages for Cubley and Biggin School using both Wirksworth and Rodsley data.

(c) Comment on the assumption in part (a). Is it reasonable?

2.4 One of four monthly-read rain gauges on a catchment area develops a fault in a month when the other three gauges record respectively 37, 43 and 51 mm. If the average annual precipitation amounts of these three gauges are 726, 752 and 840 mm respectively and of the broken gauge 694 mm, estimate the missing monthly precipitation at the latter.

2.5 Compute the average annual rainfall, in inches depth, on the catchment area shown in Fig. 1 by

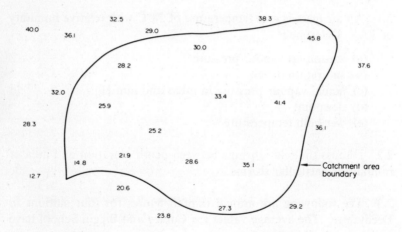

(i)   arithmetic means
(ii) Theissen method
(iii) plotting isohyets

Comment on the applicability of each method.

2.6 Discuss the setting of rain gauges on the ground and comment on the effects of wind and rain falling non-vertically on the catch.

2.7 Annual precipitation at rain gauge X and the average annual precipitation at 20 surrounding rain gauges are listed in the following table

Problems

(a) Examine the consistency of station X data.

(b) When did a change in regime occur? Discuss possible causes.

(c) Adjust the data and determine what difference this makes to the 36-year annual average precipitation at station X.

Year	Annual precipitation mm		Year	Annual precipitation mm	
	Gauge X	20 station average		Gauge X	20 station average
1972	188	264	1954	223	360
1971	185	228	1953	173	234
1970	310	386	1952	282	333
1969	295	297	1951	218	236
1968	208	284	1950	246	251
1967	287	350	1949	284	284
1966	183	236	1948	493	361
1965	304	371	1947	320	282
1964	228	234	1946	274	252
1963	216	290	1945	322	274
1962	224	282	1944	437	302
1961	203	246	1943	389	350
1960	284	264	1942	305	228
1959	295	332	1941	320	312
1958	206	231	1940	328	284
1957	269	234	1939	308	315
1956	241	231	1938	302	280
1955	284	312	1937	414	343

2.8 The data for the mean of the 20 stations in Question 2.7 should be plotted as a time series. Then plot 5-year moving averages and accumulated annual departures from the 36-year mean. Is there evidence of cyclicity or particular trends?

2.9 At a given site, a long-term wind-speed record is available for measurements at heights of 10 and 15 m above the ground. For certain calculations of evaporation, the speed at 2 m is required, so it is desired to extend the long-term record to the 2 m level. For one set of data, the speeds at 10 m and 15 m were 9.14 and 9.66 m/s respectively.

(a) what is the value of the exponent relating the two speeds and elevations?

(b) what speed would you predict for the 2 m level?

2.10   A rainfall gauge registers a fall of 9 mm in 10 minutes.

(a) how frequently would you expect such a fall at a particular place in Britain?

(b) what total volume of rain would be expected to fall on 3 km² surrounding the gauge?

2.11   What is the maximum one-day rainfall expected in Britain for a 50-year period at location X (average annual rainfall 1000 mm) and a 30-year period at location Y (average annual rainfall 1750 mm)?

2.12   What is the average rainfall over an area of 8 km² during a storm lasting 30 minutes with a frequency of once in 20 years in (a) Oxford, (b) Kumasi. Does your answer for (b) require qualification?

Chapter 3—Problems

3.1   Determine the evaporation from a free water surface using the Penman equation nomogram for the following cases

Locality	Month	Temp	h	n/D	U_2
Amsterdam (52°N)	July	18°C	0.5	0.5	1.2 m/s
Seattle (47°N)	Jan	4°C	0.8	0.3	1.5 m/s

3.2   Use the nomogram for the solution of Penman's equation to predict the daily potential evapo-transpiration from a field crop at latitude 40°N in April, under the following conditions

$$\text{mean air temp.} = 20°C \qquad \text{mean } h = 70\%$$
$$\text{sky cover} = 60\% \text{ cloud} \qquad \text{mean } U_2 = 2.5 \text{ m/s}$$
$$\text{ratio of potential evapo-transpiration to}$$
$$\text{potential evaporation} = 0\cdot7$$

3.3   Compute the potential evapo-transpiration according to Thornthwaite for two locations A and B where the local climate yields the following data

	A	B	% daylight hours of year at A		A	B	% daylight hours of year at A
Jan.	−5	−2	6	July	19	16	11
Feb.	0	2	7	Aug.	17	14	10
Mar.	5	3	$7\frac{1}{2}$	Sept.	13	10	$8\frac{1}{2}$
Apr.	9	7	$8\frac{1}{2}$	Oct.	9	8	$7\frac{1}{2}$
May	13	10	10	Nov.	5	3	7
June	17	15	11	Dec.	0	0	6

(a) at A for April (mean temp 10°C) and Nov. (mean temp = 3°C)

(b) at B for June (mean temp 20°C) and Oct. (mean temp = 8°C)

At A the average number of hours between sunrise and sunset is 13 for April and 9 for Nov. At B the figures are 14 for June and 10 for October. Use the Serra simplification for A and the nomogram for B.

3.4 Water has maximum density at 4°C; above and below this temperature its density is less. Consider a deep lake in a region where the air temperature falls below 4°C in winter.

(a) Describe what will happen in the lake in spring and autumn

(b) What will be the effect of what happens on (i) the time lag between air and water temperatures? (ii) the evaporation rate in the various seasons?

(c) Will there be a difference if the winter temperature does not drop below 4°C and if so explain why.

3.5 Calculate for location A (question 3.3) the consumptive use of water of a crop of tomatoes during a growing season from June to October if the relevant consumptive use coefficient is 10% less than that for California.

3.6 Discuss the advantages and disadvantages of evaporation pans placed above the ground surface (e.g. the U.S. Class A pan) compared to those sunk in the earth.

3.7 Draw up a water-budget for 100 units of rainfall falling on to a coniferous forest in a temperate coastal climate. Describe the processes involved and indicate the proportions of the rain that becomes involved in each.

3.8 Describe fully Penman's evaporation theory for open water surfaces. Show how each parameter used affects the evaporation and discuss how the theory differs from other evaporation formulae.

3.9 A large reservoir is located in latitude 40°30′N. Compute monthly and annual lake evaporation for the reservoir from given data using nomogram of Penman's theory. If the Class A pan evaporation at the reservoir for the year is 1143 mm, compute the pan coefficient. Assume the precipitation on the lake is as given and that the runoff represents unavoidable spillage of this precipitation during floods, what is the nett annual anticipated loss from the reservoir per square kilometre of surface in cubic metres per day?

What would the change in *evaporation* be for the month of July if the reservoir was at 40°S?

	Mean air temp. °C	Dew point °C	Average wind speed m/s	Cloud in tenths coverage	Precipita- tion mm	Runoff mm
Oct.	14·4	7·8	0·8	5·9	51	—
Nov.	8·3	1·7	1·3	7·2	99	23
Dec.	3·9	2·2	1·7	9·5	102	43
Jan.	2·2	1·9	2·1	8·7	117	58
Feb.	2·2	1·4	2·2	6·3	91	20
Mar.	4·4	1·1	1·3	5·1	69	12
Apr.	8·9	3·3	1·1	3·4	51	—
May	15	10	0·9	2·6	28	—
June	20	15·6	0·8	0·2	3	—
July	23·9	16·7	0·75	0·1	0	—
Aug.	22·8	17·8	0·7	0·0	0	—
Sept.	17·8	12·8	0·75	1·5	20	8

Chapter 4—Problems

4.1 Discuss the influence of slope of catchment and rainfall intensity on infiltration rates under constant rainfall.

4.2   Discuss the influence of forestation and agriculture on ground-water. Present arguments for and against

(a) livestock rearing
(b) crop growing
(c) forestation

on the catchment of a public water-supply reservoir.

4.3   The table below gives the hourly rainfall of three storms that gave rise to runoff equivalent of 14, 23 and 18·5 mm respectively. Determine the Φ index for the catchment.

Hour	Storm 1 mm	Storm 2 mm	Storm 3 mm
1	2	4	3
2	6	9	8
3	7	15	11
4	10	12	4
5	5	5	12
6	4		3
7	4		
8	2		

4.4   Why is the method of subtracting infiltration rates from rainfall intensities to compute hydrographs of runoff not applicable to large natural river basins?

4.5   The Antecedent Precipitation Index for a station was 53 mm on 1 October; 55 mm rain fell on 5 October, 30 mm on 7 October and 25 mm on 8 October.
Compute the API

(a) for 12 October, if $k = 0.85$
(b) for same date assuming no rain fell.

4.6   Use the co-axial relationship of Fig. 4.7 to determine how the runoff in this river changes seasonally. Assume that during week number 1 a storm of 5 inches rain lasting 72 hours occurs. Compare what happens with the effects of the same storm in week number 25, if the API in each case is 1·5 inches. Suggest which seasons of the year the weeks are in and explain why there should be a difference in runoff.

Chapter 6—Problems

6.1   A river gauging gives $Q = 4010$ m³/s. The gauging took 3 hours during which the gauge fell 0·15 m. The slope of the river surface at the gauging site at the time was 80 mm in 500 m, and the cross-section approximated a shallow rectangle 200 m wide by 11 m deep. What adjusted value of discharge would you use? What value of n in Manning's formula results?

6.2   The following discharge observations have been made on a river. Using Boyer's method adjust the figure for slope variation to produce a steady flow discharge rating curve for the river.

Gauge height ft	Measured discharge ft³/s × 1000	Rise + or Fall − ft/h
10·4	50	—
12·2	65	—
13·9	77	—
14·3	80	—
22·3	150	—
27·3	180	−0.32
28·1	228	+0·80
30·8	256	+0·525
32·6	225	−0.36
35·2	251	−0.355
38·9	338	+0·345
40·3	316	−0.22
40·8	352	+0·18
41·5	333	−0.235
42·2	362	—

6.3   Explain how observations of river discharge at particular gauge heights may be corrected, so that they fall on a smooth curve, and why this is desirable.

A river discharge was measured at $Q = 2640$ m³/s. During the 100 min of the measurement the gauge height rose from 50·40 to 50·52 m.   Level readings upstream and downstream differed by 100 mm in 700 m. The flood wave celerity was 2·2 m/s. Give the corrected rating curve co-ordinates.

6.4   An unregulated stream provides the following volumes over an 80-day period at a possible reservoir site.

(a) Plot the data in the form of a mass diagram.

(b) Determine average, maximum and minimum flow rates.

(c) What reservoir capacity would be needed to ensure maintenance of average flow for these 80 days if the reservoir is full to start with?

(d) How much water would be wasted in spillage in this case?

Day	Runoff volume $m^3 \times 10^6$	Day	Runoff volume $m^3 \times 10^6$	Day	Runoff volume $m^3 \times 10^6$
0	0	32	0·8	64	2·0
2	2·0	34	0·7	66	2·3
4	3·2	36	0·7	68	3·2
6	2·3	38	0·5	70	3·4
8	2·1	40	0·4	72	3·5
10	1·8	42	0·7	74	3·7
12	2·2	44	0·8	76	2·8
14	0·9	46	0·4	78	2·4
16	0·5	48	0·3	80	2·0
18	0·3	50	0·2		
20	0·7	52	0·2		
22	0·7	54	0·4		
24	0·6	56	0·6		
26	1·2	58	1·2		
28	0·7	60	1·4		
30	0·8	62	1·8		

6.5 The average domestic per capita demand for water in an expanding community is 0·20 m³/day. Industrial demand is 30% of total domestic requirements. The town has 100,000 inhabitants now and is expected to double its population in future.

Water is supplied from a river system with existing storage capacity of 10^7 m³ and whose mean daily discharges for each month of the year are as follows (thousands of m³)

Jan.	Feb.	Mar.	Apr.	May	June	July	Aug.	Sep.	Oct.	Nov.	Dec.
290	250	388	150	64·5	50	64·5	117	283	388	317	385

Compensation water of 1·5 m³/s is to be provided constantly.

Find, to a first approximation and for an average year, the additional storage capacity which will have to be provided if the population doubles in size. Also determine the quantity of water spilled to waste in such a year and compare it with the wastage now. Assume the existing storage is half full on 1 January.

6.6   A community of 60,000 people is increasing in size at a rate of 10% per annum. Average demand per head (for all purposes) is currently 0·20 m³/day and rising at a rate of 5% per annum. The existing water supply has a safe yield of 0·5 m³/s. A river is to be used as an additional source of supply. Its mean daily discharges for each month of the water-year are listed below in thousands of m³.

Allowing for compensation water of 3 m³/s from October–March inclusive and 5 m³/s from April–September inclusive, determine to a first approximation the storage capacity required on the river to ensure the community's water supply 20 years from now, assuming present trends continue, and that the reservoir would be full at the end of November

April 220	July	670	October	670	January	300
May 250	August	865	November	530	February	280
June 370	September	1630	December	270	March	280

6.7   List eight characteristics of drainage basins affecting their discharge hydrographs.

Chapter 7—Problems

7.1   A catchment area is undergoing a prolonged rainless period. The discharge of the stream draining it is 100 m³/s after 10 days without rain, and 50 m³/s after 40 days without rain. Derive the equation of the depletion curve and estimate the discharge after 120 days without rain.

7.2   Describe how to derive a master depletion curve for a river. What would you use it for and why?

7.3   The recession limb of a hydrograph, listed below, is to be divided into runoff and baseflow. Carry out this separation

    (a) by finding the point of discontinuity on the recession limb
    (b) by finding the depletion curve equation and extrapolating back in time.

Comment on your results.

Time h	Flow m³/s	Time h	Flow m³/s
15	41·1	33	10·0
18	35·8	36	8·3
21	25·0	39	7·0
24	19·2	42	5·8
27	15·1	45	4·9
30	12·2	48	4·1

7.4 The hydrograph tabulated below was observed for a river draining a 40 square miles catchment, following a storm lasting 3 h.

Hour	ft³/s	Hour	ft³/s	Hour	ft³/s
0	450	24	3500	48	1070
3	5500	27	3000	51	950
6	9000	30	2600	54	840
9	7500	33	2210	57	750
12	6500	36	1890	60	660
15	5600	39	1620	63	590
18	4800	42	1400	66	540
21	4100	45	1220		

Separate base-flow from runoff and calculate total runoff volume. What was the nett rainfall in inches per hour? Comment on the severity and likely frequency of such a storm in the United Kingdom.

7.5 Write down the three major principles of unit hydrograph theory illustrating their application with sketches.

Given below are three unit hydrographs derived from separate storms on a small catchment, all of which are believed to have resulted from 3 h rains. Derive the average unit hydrograph and confirm its validity if the drainage area is 5·25 square miles.

Hours	Storm 1	Storm 2	Storm 3
0	0	0	0
1	165	37	25
2	547	187	87
3	750	537	260
4	585	697	505
5	465	608	660
6	352	457	600
7	262	330	427
8	195	255	322
9	143	195	248
10	97	135	183
11	60	90	135
12	33	52	90
13	15	30	53
14	7	12	24
15	0	0	0

All values in cubic feet per second.

7.6 The 4-hour unit hydrograph for a 550 km² catchment is given below.

A uniform-intensity storm of 4 hours' duration with an intensity of 6 mm/h is followed after a 2 hour break by a further uniform-intensity storm of 2 hours' duration and an intensity of 11 mm/h. The rain loss is estimated at 1 mm/h on both storms. Base flow was estimated to be 10 m³/s at the beginning of the first storm and 40 m³/s at the end of the runoff period of the second storm.

Compute the likely peak discharge and its time of occurrence.

Hours	Q m³/s	Hours	Q m³/s
0	0	12	62
1	11	13	51
2	71	14	40
3	124	15	31
4	170	16	24
5	198	17	17
6	172	18	11
7	147	19	5
8	127	20	3
9	107	21	0
10	90		
11	76		

7.7 The 4-hour unit hydrograph for a river-gauging station draining a catchment area of 554 km², is given below. Make any checks possible on the validity of the unit graph. Find the probable peak discharge in the river, at the station from a storm covering the catchment and consisting of two consecutive 3-hour periods of nett rain of intensities 12 and 6 mm/h respectively. Assume baseflow rises linearly during the period of runoff from 30 to 70 m³/s.

Time h	Unit hydrograph m³/s	Time h	Unit hydrograph m³/s
0	0	12	62
1	11	13	51
2	60	14	39
3	120	15	31
4	170	16	23
5	198	17	16
6	184	18	11
7	153	19	6
8	127	20	3
9	107	21	0
10	91		
11	76		

7.8 A drought is ended over a catchment area of 100 km² by uniform rain of 36 mm falling for 6 hours. The relevant hydrograph of the river draining the area is given below, the rain period having been between hours 3 and 9. Use this data to predict the maximum discharge that might be expected following a 50 mm fall in 3 hours on the catchment. Qualify the forecast appropriately.

Hours	Discharge m³/s	Hours	Discharge m³/s
0	3	24	25
3	3	27	21
6	10	30	17
9	25	33	13·5
12	39	36	10·5
15	43	39	8
18	37	42	5·5
21	30·5	45	4
		48	3·9

7.9   Using the data and catchment of Question 7.7 find the probable peak discharge in the river, at the station, from a storm covering the catchment and consisting of three consecutive 2-hour periods of rain producing 7, 14 and 12 mm runoff respectively. Assume base flow rises from 10 m³/sec to 20 m³/sec during the total period of runoff.

Chapter 8—Problems

8.1   A catchment can be divided into ten sub-areas by isochrones in the manner shown in the table below, the catchment lag T_L being 10 hours.

Hour	1	2	3	4	5	6	7	8	9	10
Area in km²	14	30	84	107	121	95	70	55	35	20

A single flood recording is available from which the storage coefficient K is found as 8 hours. Derive the 2-hour unit hydrograph for the catchment.

8.2   Tabulated below is the inflow I to a river reach where the storage constants are $K = 10$ h and $x = 0$. Find graphically the outflow peak in time and magnitude.

What would be the effect of making $x > 0$?

Time h	Inflow I m³/s	Time h	Inflow I m³/s
0	28·3	40	90·6
5	26·9	45	70·8
10	24·1	50	53·8
15	62·3	55	42·5
20	133·1	60	34·0
25	172·7	65	28·3
30	152·9	70	24·1
35	121·8		

Assume outflow at hour 11 is 28·3 m³/s and starting to rise.

8.3   A storm over the catchment shown in Fig. 2 generates simultaneously at A and B the hydrograph listed below. Use the Muskingum streamflow-routing technique to determine the combined maximum discharge at C. The travel time for the mass centre of the flood between A and C is 9 hours and the factor $x = 0.33$. Any local inflow is neglected.

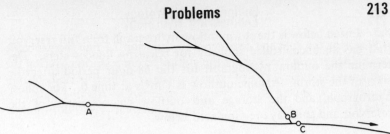

FIG. 2

Hours	Q m³/s	Hours	Q m³/s
0	10	24	91
3	35	27	69
6	96	30	54
9	163	33	41
12	204	36	33
15	210	39	27
18	190	42	24
21	129		

8.4   Define the instantaneous unit hydrograph of a catchment area, and describe how it may be used to derive the *n*-h unit graph.

A catchment area is 400 km² in total and is made up of the sub-areas bounded by the isochrones tabulated below. From a short-storm hydrograph it is known that $T_L = 9$ h and the storage coefficient $K = 5.5$ h.

Derive the 3 h unitgraph.

Sub-area bounded by isochrone h	Area km²
1	15
2	30
3	50
4	75
5	80
6	60
7	45
8	25
9	20

8.5 Listed below is the storm inflow hydrograph for a full reservoir that has an uncontrolled spillway for releasing flood waters. Determine the outflow hydrograph for the 48-hour period after the start of the storm. Assume outflow is 1 m^3/s at time 0. The inflow hydrograph, and the storage and outflow characteristics of the reservoir and spillway are tabulated below.

INFLOW HYDROGRAPH

3-h intervals	m^3/s	3-h intervals	m^3/s
0	1·5	12	54
1	156	13	45
2	255	14	40
3	212	15	34
4	184	16	28
5	158	17	23
6	136	18	17
7	116	19	11
8	99	20	8·5
9	85	21	5·5
10	74	22	3·0
11	62		

RESERVOIR CHARACTERISTICS

Height above spillway crest m	Storage $m^3 \times 10^6$	Outflow m^3/s	Height above spillway crest m	Storage $m^3 \times 10^6$	Outflow m^3/s
0·2	0·30	1·21	3·0	6·80	70·15
0·4	0·62	3·42	3·2	7·38	77·28
0·6	0·96	6·27	3·4	7·98	84·64
0·8	1·35	9·66	3·6	8·60	92·21
1·0	1·70	13·50	3·8	9·25	100·00
1·2	2·10	17·75	4·0	9·90	108·00
1·4	2·57	22·36	4·2	10·50	116·20
1·6	3·00	27·32	4·4	11·21	124·60
1·8	3·52	32·60	4·6	11·90	133·19
2·0	4·05	38·18	4·8	12·62	141·97
2·2	4·57	44·05	5·0	13·35	150·93
2·4	5·10	50·19	5·2	14·10	160·08
2·6	5·68	56·60	5·4	14·88	169·40
2·8	6·22	63·25			

Problems 215

Chapter 9—Problems

9.1 A contractor plans to build a cofferdam in a river subject to annual flooding. Hydrological records over 30 years indicate a maximum flood flow of 7800 m³/s and a minimum of 2000 m³/s. The observed annual maxima plot as a straight line on semi-logarithmic paper where return period is plotted logarithmically.

The cofferdam will be in the river during four consecutive flood seasons and it is decided to build it sufficiently high to protect against the 20-year flood.

Evaluate (without plotting) the 20-year flood and determine the probability of its occurrence during the cofferdam's life.

9.2 The annual precipitation data for Edinburgh is given below for the years 1948–1963 inclusive. From this data

(i) estimate the maximum annual rainfall that might be expected in a 20-year period and a 50-year period;

(ii) define the likelihood of the 20-year maximum being equalled or exceeded in the 9 years since 1963.

Year	Precipitation (in.)	Year	Precipitation (in.)
1948	36·37	1956	28·17
1949	28·01	1957	25·68
1950	28·88	1958	29·51
1951	30·98	1959	18·04
1952	24·41	1960	24·38
1953	23·64	1961	25·29
1954	35·15	1962	25·84
1955	18·08	1963	30·24

9.3 Discuss the methods in common use for plotting the frequency of flood discharge in rivers. List the separate steps to be taken to predict, for a particular river, the flood discharge with a probability of occurrence of 0·005 in any year. Assume 50 years of stage observations and a number of well-recorded discharge measurements with simultaneous slope observations.

9.4 The following table lists in order of magnitude the largest recorded discharge of a river with a drainage area of 12,560 km².

Date		m^3/s		Date		m^3/s
1948	29 May	2804		1918	10 June	1495
1948	22 May	2450		1929	24 May	1492
1933	10 June	2305		1943	29 May	1478
1928	26 May	2042		1922	26 May	1476
1932	14 May	2042		1919	23 May	1473
1933	4 June	2016		1936	10 April	1433
1917	17 June	1997		1936	5 May	1410
1947	8 May	1980		1923	26 May	1405
1917	30 May	1974		1927	28 April	1314
1921	20 May	1974		1939	4 May	1314
1927	8 June	1943		1934	25 April	1300
1928	9 May	1861		1945	6 May	1257
1927	17 May	1818		1935	24 May	1246
1913	15 May	1801		1920	18 May	1235
1938	19 April	1796		1914	18 May	1195
1936	15 May	1790		1931	7 May	1155
1922	6 June	1767		1911	13 June	1119
1932	21 May	1762		1940	12 May	1051
1912	20 May	1753		1942	26 May	1051
1938	28 May	1722		1946	6 May	1037
1922	19 May	1716		1926	19 April	1017
1925	20 May	1694		1937	19 May	971
1924	13 May	1668		1944	16 May	969
1917	9 June	1609		1930	25 April	878
1916	19 June	1586		1941	13 May	818
1912	21 May	1563		1915	19 May	799
1918	5 May	1495				

The mean of annual floods is 1502 m³/s and the standard deviation of the annual series is 467 m³/s.

Compute return periods and probabilities for both partial and annual series. Plot the partial series data on semi-log plotting paper and the annual series on log-normal and Gumbel probability paper. Estimate from each the discharge for a flood with a probability of once in 200 years.

9.5 The annual rainfall in inches for Woodhead Reservoir for the period of record 1921–1960 is listed below.

The mean and standard deviation for the period are 49·69 in. and 7.08 in. respectively.

Arrange the data in ranking order. Compute return periods and probability. Plot the data on probability paper.

Year	Rainfall	Year	Rainfall
1921	44·48	1941	46·79
1922	56·25	1942	43·38
1923	65·57	1943	45·87
1924	45·72	1944	59·00
1925	44·78	1945	42·74
1926	48·02	1946	55·39
1927	54·48	1947	41·04
1928	51·53	1948	45·03
1929	48·11	1949	44·11
1930	59·03	1950	52·15
1931	60·59	1951	55·43
1932	47·20	1952	48·87
1933	38·16	1953	43·26
1934	45·38	1954	63·13
1935	54·62	1955	36·46
1936	53.03	1956	56·57
1937	40·98	1957	51·79
1938	50·88	1958	51·22
1939	49·95	1959	39·58
1940	46·27	1960	60·66

(a) What are the 50-year and 100-year annual rainfalls? How do these compare with predictions made using Gumbel's theory? What qualification needs to be made if using the latter results?

(b) What is the probability that the 20-year rainfall will be exceeded in a 10-, a 20- and a 40-year period?

(c) A nearby waterworks plant is to be designed for a useful life of 50 years. It can tolerate an occasional rainfall year of 70 in. What is the probability that this amount may occur during the project's life?

General Bibliography

Chapter 2

1. MAIDENS, A. L.: New Meteorological Office rain-gauges, *The Meteorological Magazine*, Vol. 94, No. 1114, p. 142, May 1965.
2. GOODISON, C. E. and BIRD, L. G.: Telephone interrogation of rain-gauges, *Ibid.*, p. 144.
3. GREEN, M. J.: Effects of exposure on the catch of rain gauges, *T.P.* 67 *Wat. Res. Assoc.*, July 1969.
4. BLEASDALE, A.: Rain gauge networks development and design with special reference to the United Kingdom, *IASH Symposium on Design of Hydrological Networks*, Quebec, 1965.
5. BILHAM, E. G.: *The Classification of Heavy Falls of Rain in Short Periods*, H.M.S.O., London, 1962 (republished).
6. A guide for engineers to the design of storm-sewer systems. *Road Res. Lab.*, *Road Note* 35, H.M.S.O., 1963.
7. HOLLAND, D. J.: Rain intensity frequency relationships in Britain, *British Rainfall* 1961, H.M.S.O., 1967.
8. YARNALL, D. L.: Rainfall intensity-frequency data, *U.S. Dept. Agric.*, Misc. pub. 204, Washington D.C., 1935.
9. LINSLEY, R. K. and KOHLER, M. A.: Variations in storm rainfall over small areas, *Trans. Am. Geophys. Union*, Vol. 32, p. 245, April 1951.
10. HOLLAND, D. J.: The Cardington rainfall experiment, *The Meteorological Magazine*, Vol. 96, No. 1140, pp. 193–202, July 1967.
11. YOUNG, C. P.: Estimated rainfall for drainage calculations, LR 595, Road Res. Lab., H.M.S.O., 1973.
12. THIESSEN, A. H.: Precipitation for large areas, *Monthly Weather Review*, Vol. 39, pp. 1082, July 1911.

Further references

Standards for methods and records of hydrologic measurements, *Flood Control Series no.* 6, United Nations 1954.

General Bibliography   219

LANGBEIN, W. B.: Hydrologic data networks and methods of extrapolating or extending available hydrologic data, *Flood Control Series No. 15*, United Nations 1960.

Guide to Hydrometeorological Practices, *U.N. World Met. Org.*, No. 168, T.P. 82, Geneva 1965.

Chapter 3

13. PENMAN, H. L.: Natural evaporation from open water, bare soil and grass, *Proc. Roy. Soc. (London)*, A vol. 193, p. 120, April 1948.
14. THORNTHWAITE, C. W.: An approach towards a rational classification of climate, *Geographical Review*, Vol. 38, p. 55, (1948).
15. British Rainfall 1939 (and subsequent years), H.M.S.O., London.
16. LAW, F.: The aims of the catchment studies at Stocks Reservoir, Slaidburn, Yorkshire (unpublished comm. to Pennines Hydrological Group, Inst. Civ. Eng., September 1970).
17. HOUK, I. E.: *Irrigation Engineering*, Vol. 1, Wiley, New York, 1951.
18. OLIVIER, H.: *Irrigation and climate*, Arnold, London 1961.
19. THORNTHWAITE, C. W.: The moisture factor in climate, *Trans. Am. Soc. Civ. Eng.*, Vol. 27, No. 1, p. 41, Feb. 1946.
20. BLANEY, H. F. and CRIDDLE, W. D.: Determining water requirements in irrigated areas from climatological and irrigation data, *Div. Irr. and Wat. Conserv.*, *S.C.S. U.S. Dept. Agric.*, SCS-TP-96 Washington D.C., 1950.
21. BLANEY, H. F.: Definitions, methods and research data, A symposium on the consumptive use of water, *Trans. Am. Soc. Civ. Eng.*, Vol. 117, p. 949, (1952).
22. HARRIS, F. S.: The duty of water in Cache Valley, Utah, *Utah Agr. Exp. Sta. Bull.*, 173, 1920.
23. FORTIER, SAMUEL: Irrigation requirements of the arid and semi-arid lands of the Missouri and Arkansas River basins, *U.S. Dept. Agr. Tech. Bull.* 26, 1928.

Further references

HORSFALL, R. A.: Planning irrigation projects, *J. Inst. Engrs. Aust.*, 22, No. 6, June 1950.

WHITE, W. N.: A method of estimating ground water supplies based on discharge by plants and evaporation from soil. Results of investigations in Escalante Valley, Utah, *U.S. Geological Survey Water Supply*, Paper 659-A, 1932.

HILL, R. A.: Operation and maintenance of irrigation systems, Paper 2980, *Trans. Am. Soc. Civ. Eng.*, 117, p. 77, 1952.

CRIDDLE, W. D.: Consumptive use of water and irrigation requirements, *J. Soil Wat. Conserv.*, 1953.

CRIDDLE, W. D.: Methods of computing consumptive use of water, Paper 1507, *Proc. Amer. Soc. Civ. Eng.*, **84**, Jan. 1958.

ROHWER, CARL: Evaporation from different types of pans, *Trans. Am. Soc. Civ. Eng.*, Vol. 99, p. 673, 1934.

HICKOX, G. H.: Evaporation from a free water surface, *Trans. Amer. Soc. Civ. Eng.*, Vol. 111, Paper 2266, 1946, and discussion by C. Rohwer.

LOWRY, R. and JOHNSON, A. R.: Consumptive use of water for agriculture, *Trans. Amer. Soc. Civ. Engrs.*, Vol. 107, paper 2158, 1942, and discussion by Rule, R. E., Foster, E. E., Blaney, H. F. and Davenport, R. W.

FORTIER, S. and YOUNG, A. A.: Various articles in *Bull. U.S. Dep. Agric.* Nos. 1340 (1925), 185 (1930), 200 (1930), 379 (1933).

PENMAN, H. L.: Estimating evaporation, *Trans. Amer. Geophys. Union*, Vol. 31, p. 43, Feb. 1956.

Chapter 4

24. NASSIF, S. H. and WILSON, E. M.: The influence of slope and rain intensity on runoff and infiltration, *Bull. I.A.H.S.*, Vol. 20, No. 4, 1976.

25. HORTON, R. E.: The role of infiltration in the hydrologic cycle, *Trans. Am. Geophys. Union*, Vol. 14, pp. 443–460, 1933.

26. BOUCHARDEAU, A. and RODIER, J.: Nouvelle méthode de détermination de la capacité d'absorption en terrains perméables, *La Houille Blanche*, No. A, pp. 531–526, July/Aug. 1960.

27. SOR, K. and BERTRAND, A. R.: Effects of rainfall energy on the permeability of soils, *Proc. Am. Soc. Soil Sci.*, Vol 26, No. 3, 1962.

28. HORTON, R. E.: Determination of infiltration capacity for large drainage basins, *Trans. Am. Geophys. Union*, Vol. 18, p. 371, 1937.

29. SHERMAN, L. K.: Comparison of F-curves derived by the methods of Sharp and Holtan and of Sherman and Mayer, *Trans. Am. Geophys. Union*, Vol. 24 (2), p. 465, 1943.

30. BUTLER, S. S.: *Engineering Hydrology*, Prentice-Hall, Englewood Cliffs, 1957.

31. LINSLEY, R. K., KOHLER, M. A. and PAULHUS, J. L. H.: *Hydrology for Engineers*, (p. 162), McGraw-Hill, New York, 1958.

32. Estimated Soil Moisture deficit over Gt. Britain: *Explanatory Notes Meteorological Office, Bracknell* (issued twice monthly).

33. PENMAN, H. L.: The dependence of transpiration on weather and soil conditions, *Journal of Soil Science*, Vol. 1, p. 74, 1949.

34. GRINDLEY, J.: Estimation of soil moisture deficits, *Meteorological Magazine*, Vol. 96, p. 97, 1967.

Further references

BELL, J. P.: Neutron probe practice, Institute of Hydrology *Report No. 19*, Wallingford, U.K.

HORTON, R. E.: Analyses of runoff-plot experiments with varying infiltration capacity, *Trans. Amer. Geophys. Union*, Pt. IV, p. 693, 1939.

WILM, H. G.: Methods for the measurements of infiltration, *Trans. Amer. Geophys. Union*, Pt. III, p. 678, 1941.

Chapter 5

WENZEL, L. K.: Methods for determining permeability of water bearing materials, *U.S. Geol. Surv. Water Supply*, Paper 887, 1942.

KIRKHAM, DON: Measurement of the hydraulic conductivity of soil in place, Symposium on Permeability of Soils, *Am. Soc. Testing Materials*, special Tech. Publ. 163, p. 80, 1955.

CHILDS, E. C. and COLLIS-GEORGE, N.: The permeability of porous materials, *Proc. Roy. Soc.*, A201: p. 392, 1950.

ARONOVICI, V. S.: The mechanical analysis as an index of subsoil permeability, *Proc. Am. Soc. Soil Sci.*, Vol. 11, p. 137, 1947.

TODD, DAVID K.: *Ground Water Hydrology*, John Wiley, New York, 1959.

HUISMAN, L.: *Groundwater Recovery*, Macmillan, London, 1972.

VERRUIJT, A.: *Theory of Groundwater Flow*, Macmillan, London, 1970.

CEDERGREEN, H. R.: *Seepage, Drainage and Flow Nets*, John Wiley, New York, 1967.

Chapter 6

35. HOSEGOOD, P. H. and BRIDLE, M. K.: A feasibility study and development programme for continuous dilution gauging, Institute of Hydrology, *Report No.* 6, Wallingford, U.K.

36. ISO/R. 55, 1966, Liquid flow measurement in open channels; dilution methods for measurement of steady flow, Part 1, constant rate injection.

37. CORBETT, DON M. and others, Stream-Gaging Procedure, Water Supply Paper 888, *U.S. Geol. Survey*, Washington D.C., 1943.

38. BOYER, M. C.: Determining Discharge at Gaging Stations affected by variable slope, *Civil Eng.*, Vol. 9, p. 556, 1939.

39. MITCHELL, W. D.: Stage-Fall-Discharge Relations for Steady flow in prismatic channels, *U.S. Geol. Survey*, Water Supply Paper 1164, Washington D.C., 1954.

40. STEVENS, J. C.: A method of estimating stream discharge from a limited number of gagings, *Eng. News*, July 18, 1907.

41. KOELZER, V. A.: *Reservoir Hydraulics*, Sect. 4, Handbook of Applied Hydraulics, ed. by Davis and Sorenson, 3rd Edition, McGraw-Hill, New York, 1969.

42. JENNINGS, A. H.: World's greatest observed point rainfalls, *Monthly Weather Review*, Vol. 78, p. 4, Jan. 1950.

Further references

B.S. 3680, Part 3, 1964; Part 4, 1965.

Logarithmic plotting of stage-discharge observations, Tech. Note 3. *Water Resources Board*, Reading, 1966.

HORTON, R. E.: Erosional development of streams and their drainage basins, *Bull. Geol. Soc. Am.*, Vol. 56, p. 275, March 1945.

STRAHLER, Statistical analysis of geomorphic research, *Journ. of Geol.*, Vol. 62, No. 1, 1964.

ACKERS, P. and HARRISON, A. J. M.: Critical depth flumes for flow measurement in open channels, *Hyd. Res. Paper No. 5*, London, H.M.S.O., 1963.

PARSHALL, R. L.: Measuring water in irrigation channels with Parshall flumes and small weirs, *U.S. Dept. Agric. Circ.* 843, 1950.

ACKERS, P.: Flow measurement by weirs and flumes, *Int. Conf. on Mod. Dev. in Flow Measurement*, Harwell, 1971, Paper No. 3.

WHITE, W. R.: Flat-vee weirs in alluvial channels, *Proc. A.S.C.E.*, 97 HY3, pp. 395–408, March 1971.

WHITE, W. R.: The performance of two dimensional and flat-V triangular profile weirs, *Proc. I.C.E.*, Supplement (ii), pp. 21–48, 1971.

BURGESS, J. S. and WHITE, W. R.: Triangular profile (Crump) weir: two dimensional study of discharge characteristics, Rpt. No. INT 52, H.R.S. Wallingford, 1952.

HARRISON, A. J. M. and OWEN, M. W.: A new type of structure for flow measurement in steep streams, *Proc. I.C.E.*, 36, pp. 273–296, 1967.

SMITH, C. D.: Open channel water measurement with the broad-crested weir, *Int. Comm. on Irrig. and Drainage Bull.*, 1958, pp. 46–51.

Chapter 7

43. SHERMAN, L. K.: Stream flow from rainfall by the unitgraph method, *Eng. News Record*, Vol. 108, p. 501, 1932.
44. BERNARD, M.: An approach to determinate stream flow, *Trans. Am. Soc. Civ. Eng.*, Vol. 100, p. 347, 1935.
45. LINSLEY, R. K., KOHLER, M. A. and PAULHUS, J. L. H., *Applied Hydrology*, pp. 448–49, McGraw-Hill, New York, 1949.
46. COLLINS, W. T.: Runoff distribution graphs from precipitation occurring in more than one time unit, *Civil Engineering*, Vol. 9, No. 9, p. 559, Sept. 1939.
47. SNYDER, F. F.: Synthetic unitgraphs, *Trans. Am. Geophys. Union*, 19th Ann. meeting 1938, Pt. 2, p. 447.
48. HURSH, C. R.: Discussion on Report of the committee on absorption and transpiration, *Trans. Am. Geophys. Union*, 17th Ann. meeting, 1936, p. 296.
49. SNYDER, F. F.: Discussion on ref. 47.
50. LINSLEY, R. K.: Application of the synthetic unitgraph in the western mountain States, *Trans. Am. Geophys. Union*, 24th Ann. Meeting, 1943, Pt. 2, p. 580.
51. TAYLOR, A. B. and SCHWARZ, H. E.: Unit hydrograph lag and peak flow related to basin characteristics, *Trans. Am. Geophys. Union*, Vol. 33, p. 235, 1952.

Further references

BARNES, B. S.: Consistency in unitgraphs, *Proc. Am. Soc. Civ. Eng.*, 85, HY8, p. 39, Aug. 1959.

MORRIS, W. V : Conversion of storm rainfall to runoff, *Proc. Symposium No. 1, Spillway Design Floods*, N.R.C., Ottawa, p. 172, 1961.

MORGAN, P. E. and JOHNSON, S. M.: Analysis of synthetic unitgraph methods, *Proc. Amer. Soc. Civ. Eng.*, 88 HY5, p. 199, Sept. 1962.

BUIL, J. A.: Unitgraphs for non uniform rainfall distribution, *Proc. Am. Soc. Civ. Eng.*, 94, HY1, p. 235, Jan. 1968.

Chapter 8

52. McCARTHY, G. T.: The unit hydrograph and flood routing, unpublished paper presented at the Conference of the North Atlantic Division, Corps of Engineers, U.S. Army, New London, Conn. June 24, 1938. Printed by U.S. Engr. Office, Providence R.I.

53. CARTER, R. W. and GODFREY, R. G.: Storage and Flood Routing, *U.S. Geol. Survey Water-Supply*, Paper 1543-B, p. 93, (1960).

54. WILSON, W. T.: A graphical flood routing method, *Trans. Am. Geophys. Union*, Vol. 21, part 3, p. 893, 1941.

55. KOHLER, M. A.: Mechanical analogs aid graphical flood routing, *J. Hydraulics Div.*, ASCE 84, April 1958.

56. LAWLER, E. A.: Flood routing, Sec. 25-II, *Handbook of Applied Hydrology*, ed. Ven Te Chow, McGraw-Hill, New York, 1964.

57. CLARK, C. O.: Storage and the unit hydrograph, *Trans. Amer. Soc. Civ. Engr.*, Vol. 110, p. 1419, (1945).

58. O'KELLY, J. J.: The employment of unit hydrographs to determine the flows of Irish arterial drainage channels, *Proc. Instn. Civ. Engrs.*, Pt. III, Vol. 4, p. 365, (1955).

59. NASH, J. E.: Determining runoff from rainfall, *Proc. Instn. Civ. Engrs.*, Vol. 10, p. 163, (1958).

60. NASH, J. E.: Systematic determination of unit hydrograph parameters, *Journ. Geophys. Res.*, Vol. 64, p. 111, (1959).

61. NASH, J. E.: A unit hydrograph study, with particular reference to British catchments, *Proc. Instn. Civ. Engrs.*, Vol. 17, p. 249, (1960).

62. VEN TE CHOW, *Handbook of Applied Hydrology*, Sect. 14, McGraw Hill, New York, (1964).

Chapter 9

63. MORGAN, H. D.: Estimation of design floods in Scotland and Wales, Paper No. 3, Symposium on River Flood Hydrology, *Instn. Civ. Engrs.*, London, 1966.

64. *Flow in California streams*, Calif. Dept. Public Works, Bull. 5, 1923.

65. HAZEN, A.: *Flood Flow*, Wiley, New York, 1930.

66. HAZEN, A.: Storage to be provided in impounding reservoirs for municipal water supply, *Trans. A.S.C.E.*, Vol. 77, p. 1539, 1914.

67. WHIPPLE, G. C.: The element of chance in sanitation, *J. Franklin Inst.*, Vol. 182, p. 37, et seq., 1916.

68. GUMBEL, E. J.: On the plotting of flood discharges, *Trans. Am. Geophys. Union*, Vol. 24, Pt. 2, p. 699, 1943.
69. GUMBEL, E. J.: Statistical theory of extreme values and some practical applications, *Natl. Bur. Standards (U.S.) Appl. Math. Ser.*, 33, Feb. 1954.
70. POWELL, R. W.: A simple method of estimating flood frequency, *Civil Eng.*, Vol. 13, p. 105, 1943.
71. *Ibid.*, discussion by E. J. Gumbel, p. 438.
72. VEN TE CHOW and YEVJEVICH, V. M.: *Statistical and Probability Applied Hydrology*, ed. Ven Te Chow, McGraw-Hill, New York, 1964.
73. DALRYMPLE, TATE: Flood Frequency Analysis, *U.S. Geol. Water Supply*, Paper 1543-A, (1960).
74. PAULHUS, J. L. H. and GILMAN, C. S.: Evaluation of probable maximum precipitation, *Trans. Am. Geophys. Union*, 34, p. 701, 1953.
75. Generalised estimates of probable maximum precipitation over the U.S. east of the 105th meridian, Hydrometeorological Report No. 23, *U.S. Weather Bureau*, Washington, 1947.
76. Generalised estimates of probable maximum precipitation of the United States west of the 105th meridian for areas to 400 square miles and durations to 24 hours. Tech. Paper 38, *U.S. Weather Bureau*, Washington, 1960.
77. Manual for depth duration area analysis of storm precipitation, *U.S. Weather Bureau Co-operative Studies Tech. Paper*, No. 1, Washington, 1946.
78. HERSHFIELD, D. M.: Estimating the probable maximum precipitation, *Proc. Am. Soc. Civ. Eng.*, 87, p. 99, September 1961.
79. *Flood Study Report*, Institute of Hydrology, Wallingford, U.K., 1975.
80. BLEASDALE, A.: The distribution of exceptionally heavy daily falls of rain in the United Kingdom, *Journ. Inst. Wat. Eng.*, Vol. 17, p. 45, Feb. 1963.
81. WIESNER, C. J.: Analysis of Australian storms for depth, duration, area data, *Rain Seminar, Commonwealth Bureau of Meteorology*, Melbourne, 1960.
82. WOLF, P. O.: Comparison of methods of flood estimation, *Symposium on River Flood Hydrology, Instn. Civ. Engrs.*, London, 1966 and discussion by T. O'Donnell.
83. PETERSON, K. R.: A precipitable water nomogram, *Bull. Amer. Met. Soc.*, 42, p. 199, 1961.
84. SOLOT, S.: Computation of depth of precipitable water in a column of air, *Mon. Weath. Rev.*, 67, p. 100, 1939.
85. BINNIE, G. M. and MANSELL-MOULLIN, M.: The estimated probable maximum storm and flood on the Jhelum River—a tributary of the Indus, Paper No. 9 Symposium on River Flood Hydrology, *Instn. of Civ. Engrs.*, London, 1966.
86. *Handbook of Meteorology*, ed. by Berry, Bollay and Beers, p. 1024, McGraw-Hill, New York, 1949.

87. PAULHUS, J. L. H.: Indian ocean and Taiwan rainfalls set new records, *Mon. Weather Rev.*, 93, p. 331, May 1965.

88. BLEASDALE, A.: Private communication to the author, *Met. Office, Bracknell*, May 1968.

89. BROOKS, C. E. P. and CARRUTHERS, N.: *Handbook of Statistical methods in meteorology*, H.M.S.O., p. 330, London, 1953.

90. COCHRANE, N. J.: Lake Nyasa and the River Shire, *Proc. I.C.E.*, Vol. 8, p. 363, 1957.

91. COCHRANE, N. J.: Possible non-random aspects of the availability of water for crops, Paper No. 3, *Conf. Civ. Eng. Problems Overseas Inst. C.E.*, London, June 1964.

Further references

LANGBEIN, W. B.: Annual floods and the partial duration flood series, *Trans. Am. Geophys. Union*, Vol. 30, p. 879, Dec. 1949.

WIESNER, C. J.: Hydrometeorology and river flood estimation, *Proc. Instn. Civ. Engrs.*, Vol. 27, p. 153, 1964.

Symposium on Hydrology of Spillway Design by the Task Force on Spillway Design Floods of the Committee on Hydrology, *Proc. Am. Soc. Civ. Eng.*, Vol. 90, HY3, May 1964.

ALEXANDER, G. N.: Some aspects of time series in hydrology, *J. Instn. Engrs. Australia*, Vol. 26, pp. 188–198, 1954.

ANDERSON, R. L.: Distribution of the serial correlation coefficient, *Ann. Math. Stat.*, Vol. 13, pp. 1–13, 1941.

BEARD, L. R.: Simulation of daily streamflow. *Proc. Internat. Hydrol. Symposium*, Fort Collins, Colorado, 6–8 Sept. 1967, Vol. 1, pp. 624–632.

FIERING, M. B.: *Streamflow Synthesis*, Macmillan, London, 1967.

HANNAN, E. J.: *Time Series Analysis*, Methuen, London, 1960.

KISIEL, C. C.: Time series analysis of hydrologic data, in *Advances in Hydroscience*, Vol. 5, pp. 1–119, ed. Ven Te Chow, Academic Press, New York, 1969.

MATALAS, N. C.: Time series analysis, *Water Resources Res.*, Vol. 3, pp. 817–29, 1967.

MATALAS, N. C.: Mathematical assessment of synthetic hydrology, *Water Resources Res.*, Vol. 3, pp. 937–45, 1967.

MORAN, P. A. P.: *An Introduction to Probability Theory*, Clarendon Press, Oxford, 1968.

O'DONNELL, T.: Computer evaluation of catchment behaviour and parameters significant in flood hydrology, Symp. on River Flood Hydrology, Inst. Civ. Eng., 1965.

QUIMPO, R. G.: Stochastic analysis of daily river flows, *Proc. Am. Soc. Civ. Engrs, J. Hydraul. Div.*, Vol. 94, HY1, pp. 43–57, 1968.

YEVJEVICH, V. M. and JENG, R. I.: Properties of non-homogeneous hydrologic series, Colorado State University, Hydrology Paper No. 32, 1969.

Answers to Problems

Chapter 2

2.1 (a) 28·32 mm Hg; (b) 8·50 mm Hg; (c) 19·82 mm Hg or 26·96 mb; (d) 22·0 °C; (e) 23·7 °C

2.3 (a) 29·99 in; (b) Cubley 29·78 in; (c) Biggin 42·0 in

2.4 39 mm

2.5 29·8 in; 31·2 in; 27·0 in

2.7 (b) about 1951; (c) assuming earlier period correct, increases it from 279 to 330 mm/yr

2.9 (a) 0·136; (b) 7·34 m/s

2.10 (a) once in 4 years; (b) $23·8 \times 10^3$ m³

2.11 at X, 75 mm; at Y, 92 mm

2.12 (a) 19 mm Oxford; (b) 65 mm Kumasi (if equation 2.12 derived for United Kingdom applies in Ghana)

Chapter 3

3.1 Amsterdam 3·8 mm/day; Seattle 0·3 mm/day

3.2 2·5 mm/day

3.3 (a) April 5·24 cm; Nov 1·01 cm; (b) Jun 11·67 cm; Oct 3·70 cm

3.5 17·6 in

3.9 Annual E_0 = 950 mm; pan coefficient 0·83; surface loss = $1·32 \times 10^3$ m³/km²/day; change = 142 mm less

Chapter 4

4.3 4·1 mm

4.5 (a) 53 mm; (b) 9 mm

Chapter 6

6.1 4085 m³/s; $n = 0·032$

6.3 2560 m³/s at 50·46 m

6.4 (b) 0·705, 1·85 and 0·1 million m³/day; (c) 18·0 × 10⁶ m³; (d) 5·0 × 10⁶ m³

6.5 Additional storage 3·65 × 10⁶ m³; future spillage = 11·9 × 10⁶ m³; present spillage = 20·8 × 10⁶ m³

6.6 45 × 10⁶ m³

Chapter 7

7.1 8 m³/s

7.3 Point N at hour 33

7.4 Point N at 36 h; 1·61 in/h; extremely severe; frequency probably < once in 100 yr

7.5 peak of u.h. at 710 ft³/s and 4·2 h

7.6 about 686 m³/s at 9 h

7.7 923 m³/s at 6 h

7.8 76 m³/s assuming same Φ index and baseflow of 4 m³/s

7.9 614 m³/s at 7 h

Chapter 8

8.1 At 2 h intervals: 0, 6·95, 42·1, 95·6, 126·6, 125·9, 106·3, etc.

8.2 132 m³/s at 35 h

8.3 353 m³/s

8.4 At 3 h intervals: 0, 19·8, 72·7, 100·2, 74·7, etc.

Chapter 9

9.1 7 060 m³/s; probability 18·6%

9.2 (i) 37·0 in; 39·3 in; (ii) 37%

9.4 About 3 220 m³/s

9.5 (a) Gumbel $P_{50} = 68·04$ in, $P_{100} = 71·91$ in; (b) 0·401; 0·642; 0·871; (c) 0·395

Index

70 12